THE POLITICS OF DISPLACEMENT

Racial and Ethnic Transition
in Three American Cities

This is a volume in the

Institute for Research on Poverty Monograph Series

A complete list of titles in this series appears at the end of this volume.

THE POLITICS OF DISPLACEMENT

Racial and Ethnic Transition in Three American Cities

Peter K. Eisinger

Institute for Research on Poverty
University of Wisconsin–Madison
Madison, Wisconsin

ACADEMIC PRESS 1980
A Subsidiary of Harcourt Brace Jovanovich, Publishers
New York   London   Toronto   Sydney   San Francisco

This book is one of a series sponsored by the Institute for Research on Poverty
of the University of Wisconsin pursuant to the provisions of the Economic
Opportunity Act of 1964.

ACADEMIC PRESS, INC.
111 Fifth Avenue, New York, New York 10003

United Kingdom Edition published by
ACADEMIC PRESS, INC. (LONDON) LTD.
24/28 Oval Road, London NW1 7DX

Library of Congress Cataloging in Publication Data

Eisinger, Peter K
    The politics of displacement.

    (Institute for Research on Poverty monograph
series)
    Bibliography: p.
    1. Afro—Americans——Michigan——Detroit——Politics
and suffrage. 2. Afro—Americans——Georgia——Atlanta
——Politics and suffrage. 3. Irish Americans——
Massachusetts——Boston——Politics and government.
4. Detroit——Politics and government. 5. Atlanta
——Politics and government. 6. Boston——Politics
and government. I. Title. II. Series: Wisconsin.
University——Madison. Institute for Research on Poverty.
Monograph series.
F574.D49N434   306'.2   80—12927
ISBN 0—12—235560—1

PRINTED IN THE UNITED STATES OF AMERICA

80 81 82 83   9 8 7 6 5 4 3 2 1

The Institute for Research on Poverty is a national center for research established at the University of Wisconsin in 1966 by a grant from the Office of Economic Opportunity. Its primary objective is to foster basic, multidisciplinary research into the nature and causes of poverty and means to combat it.

In addition to increasing the basic knowledge from which policies aimed at the elimination of poverty can be shaped, the Institute strives to carry analysis beyond the formulation and testing of fundamental generalizations to the development and assessment of relevant policy alternatives.

The Institute endeavors to bring together scholars of the highest caliber whose primary research efforts are focused on the problem of poverty, the distribution of income, and the analysis and evaluation of social policy, offering staff members wide opportunity for interchange of ideas, maximum freedom for research into basic questions about poverty and social policy, and dissemination of their findings.

To my family,
Erica, Jesse, and Sarah

Contents

List of Figures and Tables

List of Abbreviations

AC	*Atlanta Constitution*
BET	*Boston Evening Transcript*
DFP	*Detroit Free Press*
DN	*Detroit News*
NYT	*New York Times*

Foreword

In 1964, there were only 70 elected black officials at all levels of government throughout the United States. The civil rights movement and government responses, including especially the Community Action Programs of the War on Poverty, enhanced black political power. Today there are over 4600 elected black officials in the United States. Nowhere is the new political strength of blacks more apparent than in municipal government. More than 170 cities including Atlanta, Detroit, Gary, Los Angeles, Newark, and New Orleans now have black mayors. In *The Politics of Displacement*, Peter Eisinger examines the responses of the "displaced" white elites in Atlanta and Detroit to the black takeover of city hall and compares it to the Yankee response in Boston at the turn of the century to the political ascendancy of the Irish.

While studies of losers are rare, Eisinger argues that they are worthwhile because the newly victorious group's ability to govern and the well-being of their cities will depend upon the response of the old elite, and because from such studies we can learn something about ethnoracial conflict and cooperation in America. Eisinger identifies five possible responses of the elites; he labels them cooperation, maintenance, subversion, contestation, and withdrawal. In each case, he finds, cooperation was the dominant response of the old elites.

note 5

In other societies, transitions of power from one previously dominant ethnic group to another previously subservient group have been marked by violence. In our own society, ethnic and racial competition have also frequently been violent. Once the possibility of electoral takeover became serious in Boston, Atlanta, and Detroit, however, Eisinger notes that the competition for and eventually the transition of power became peaceful and orderly.

Eisinger identifies numerous reasons for the cooperative response of the old elite to electoral defeat. These include the large economic stakes of the old elites in the cities, the dependence of the new elite on cooperation from the old elite and their consequent conciliatory approach, and most important, argues Eisinger, the limits Americans place on the use of politics. Electoral victory does not in the American context set the stage for radical transformations, but rather gradual changes.

Whether the increases in political power achieved by black elites lead gradually to economic gains for the black masses is a question which is touched upon in this book. Eisinger notes that the Irish are now the second most successful white ethnic group in the United States and suggests that their political ascendancy 50 years ago is probably related to their current economic success. He also notes that black mayors in Detroit and Atlanta have begun shifting employment towards blacks. But these are only clues. Eisinger is now engaged in an intensive examination of this question. When he completes this work we will be better able to judge the efficacy of political strategies for reducing poverty.

This volume is Peter Eisinger's second in the Poverty Institute monograph series. *Patterns of Interracial Politics: Conflict and Cooperation in the City* was published in 1976. Taken together, the two volumes and Eisinger's current work represent an impressive corpus of work on interracial politics in the United States.

Irwin Garfinkel
Director
Institute for Research on Poverty

Preface

Since 1967 more than 170 towns and cities across the nation have elected black men and women as mayors. Although in numerical terms black-run cities are increasingly commonplace, the accession of blacks to urban mayoralties is, in fact, no ordinary political event. Every initial black victory in a major city (with the exception of Washington, D.C.) has been won over a white opponent. Racial considerations have generally dominated these campaigns, overwhelming other potential bases of conflict. In most cases blacks have been carried to victory on the strength of an emergent black electoral majority. What we have in effect witnessed in most of these places in the 1970s is a transition to black rule and the concomitant political displacement of whites and white-defined interests.

How whites have dealt with this transformation of their local political world is the principal concern of the present study. In particular, this book is an investigation of the adjustments to black rule made by white elites in Detroit and Atlanta since 1973 and the consequences of those adjustments for the black mayors in those places. To cast historical light upon the experience of ethnic and racial political transition I have also included a case study of Yankee adjustment to Irish rule in Boston at the turn of the century.

In a narrow sense this is a study that focuses on losers—members of groups displaced from political power by other groups they once traditionally dominated—and on their psychological and strategic adaptations. At a more general level the inquiry offers a perspective on the role of race and ethnicity in American cities and certain perhaps surprisingly adaptive qualities of American politics.

When this study began in 1974, shortly after black mayors had been elected in Detroit and Atlanta, one might have expected that the book would end up as an account of the death of two American cities. The possibilities for crippling the new administrations in those places were abundant, and many observers fully expected them to be realized. Among other scenarios, many anticipated the hasty withdrawal of white civic and economic resources from these cities. If whites were to withdraw their wealth, their industries and shops, their expertise and experience in government administration, and their prestige, they would in all likelihood leave the newly elected mayors in charge of paper kingdoms. Unable to generate substitute resources of sufficient magnitude, the black mayors would preside helplessly over the withering of the central city economy, the bankruptcy of city government, and the disintegration of public services.

The potential withdrawal of white resources from the city was not the only possibility that black mayors and those who observed their cities had to anticipate. Perhaps the white community, unaccustomed to and frightened of minority status in a black majority city, would use its resources for opposition purposes. Would whites seek to regain the mayoralty, and, if so, what means would they use? Or would they seek in some way to subvert or obstruct the black administration?

The question of what politically displaced whites do with their considerable resources in black-ruled cities is not a trivial one. American cities (it is hardly news) are in trouble. The help the federal and state governments provide these days, though steadily increasing, is not enough. For better or worse, cities must rely on the commitment of their private-sector civic and economic elites to supplement their political and administrative leadership, to maintain the local economy, to attract investment, and to build and rebuild the city. The dependence of the public sector on the private sector takes on a special significance where the two realms are controlled by people of different races and, by implication, of different levels of aggregate socioeconomic achievement. To win control of the formal apparatus of government is not enough to make possible the effective pursuit of the governing process. It is also necessary to be able to draw, without force, from the pool of resources controlled by those outside of government.

Contrary to general expectations, the transition to black rule in Atlanta and Detroit has been remarkably benign. Neither city seems to be dying in any sense. For the most part white resources have neither been withdrawn nor turned to opposition. The politics of these cities, at least into 1979, have been marked not by racial conflict but by patterns of coalition-building, cooperation, and accommodation that crossed the deep racial divide. To describe the emergence of these patterns and to understand why they have occurred rather than the more frightening possibilities that many anticipated are the purposes of this book.

The conclusions I have reached regarding the nature of the white response to black rule and its larger meaning may strike some as unusually optimistic. Yet a certain optimism is warranted. What has occurred in Atlanta and Detroit in the 1970s is particularly noteworthy when it is set against the history of race relations in those two cities themselves, against the habits of racial oppression in American society in general, and indeed against a virtually worldwide tendency to deal with ethno-racial political competition by violent means.

There are, of course, sharply drawn limits to the optimistic perspective. The relatively amicable transition to black rule does not mean that the two cities have achieved racial equality in the marketplace or freedom from racial tensions. They have not. But to concede this is not to minimize what has occurred in those places. What people in Detroit and Atlanta have accomplished during the transition from a white to a black majority is a significant demonstration of the adaptive capacities of American politics, of the possibilities for bridging the racial cleavage in American life, and of the strength of will in these cities to survive as spiritually and socially vital urban centers.

ACKNOWLEDGMENTS

Many people have contributed in one way or another to this study. My first debt of gratitude is to those politicians and officials, business people, bankers, lawyers, union officials, and others in Detroit and Atlanta who agreed to be interviewed for this study. They spoke openly, articulately, and at length with me. Their observations and opinions are treated anonymously in the text of the book, but I have listed their names in Appendix A in recognition of the help they gave me. I also wish to thank Professor Diane Fowlkes and the Georgia State University Department of Political Science for providing me with office space in Atlanta in 1975.

A number of colleagues have spent time talking with me about this study or have read and commented on various parts or all of the manuscript. These include Henry Pratt, John Armstrong, Crawford Young, Murray Edelman, David Greenstone, and Walter Jones. I am grateful also for the occasional advice of my wife, Erica Eisinger, which has saved me from what I would like to think are quite uncharacteristic errors of judgment, and to my parents, Chester and Marjorie Eisinger, for their comments and encouragement.

Darryl Solochek, Will Sullivan, and Walter Jones did a considerable amount of legwork and research assistance for me during this study, and Diane Rottenberg collated and checked the References. They made my task easier and I am grateful to them. Typists at the Institute for Research on Poverty and the Department of Political Science at the University of Wisconsin have cooperated in producing innumerable drafts for me, and Jan Blakeslee and Judith Kirkwood ably guided the book through the editorial and production process.

I wish finally to thank the Institute for Research on Poverty and its director, Irwin Garfinkel, for generous financial assistance and moral support. The Institute must not, of course, be held responsible for anything I have said in this book, although it should certainly share credit for whatever merit may be found herein.

1

The Problem of Displacement

Through the first quarter of the nineteenth century, Quincy, Massachusetts, was a small farming community populated almost entirely by people of English blood. During the 1820s, however, the ethnic character of the village began to change. The construction of a small railroad south of Boston had made granite quarrying near Quincy an economically attractive venture, and the development of this industry brought large numbers of Catholic Irish laborers to the village. By 1840 the Irish had achieved sufficient mobilization and numbers to be counted among the several contending groups seeking mastery of the Quincy town meeting (Solomon, 1956, pp. 28–32).

In 1874, with the municipal debt running high, the Adams brothers, John Quincy and Charles Francis, assumed control over their ancestral hometown and ruled the town meeting for nearly 14 years. By the 1880s, however, the Irish were beginning to assert their political strength throughout Massachusetts. They established dominance in Quincy with the help of the Knights of Labor, defeating the Adams brothers and their Anglo-Saxon allies, and driving the former out of politics entirely. When Charles Francis Adams suffered the additional personal humiliation of losing his job as president of Union Pacific Railroad in 1890, he retreated

1

from Quincy to resettle in the rural Yankee village of Lincoln, Massachusetts, a place "still safe from the unbearable vulgarities of an urban municipality. . . . For the rest of his life [Adams] never forgot that the Irish . . . had displaced him [Solomon, 1956, p. 30]."

Adams's "aristocratic recoil," as Arthur Mann (1954, p. 6) characterizes one of the major Yankee responses to the rapid social changes at the end of the nineteenth century, was not simply a reaction to personal defeat. Rather it emerged from the broader recognition of the disintegration of Yankee political and cultural hegemony in New England as a consequence in large measure of the rise of the Irish. Adams's flight from Quincy stood for the wider loss of dominance by a particular class and ethnic group. He was simply one of many losers, unaccustomed as a class to the experience of minority status, in the most dramatic case of ethnic political transition in American history.

The political displacement of the Yankee elite by the Irish in the towns and cities of the Northeast has an important parallel in contemporary American cities where new black majorities have managed to elect black mayors. Since the first important black successes in 1967 in Cleveland and Gary, blacks have won control of city hall in places as diverse as Atlanta, New Orleans, Detroit, Newark, and Los Angeles, to name the largest of the more than 170 cities with black mayors. In towns and cities where there is a growing black majority and a diminishing white minority, black control of the formal institutions of municipal government seems a likely possibility for the long run.

In the case of Gary, at least, there is modest evidence that significant segments of the white community initially refused to accept the loss of white control as a legitimate outcome of majoritarian democracy. The Eastern European ethnics, who had long controlled the Democratic machine in this prototypical company town, did not, however, withdraw to a rural idyll as Charles Francis Adams had done 80 years before: Their response was to resist and subvert the new regime. As one participant in the transition process in Gary has written (Greer, 1971):

> When the new administration took over City Hall in January 1968 it found itself without the keys to offices, with many vital records missing (for example, the file on the United States Steel Corporation in the Controller's office) and with a large part of the City government's movable equipment stolen [p. 36].

Once the initial period of transition had passed, resistance to Mayor Richard Hatcher centered in the city bureaucracy where, according to Greer, inertia, obstruction, and simple sabotage effectively blocked mayoral initiatives.

The situation in Newark, however, also a black majority city, appears to have differed from that in Gary. By the time Kenneth Gibson ran (successfully) for a third term in city hall in 1978, relations with at least the white business community were very good. Communication between city hall and the corporate boardrooms was regular and amicable. Construction in the downtown had boomed during Gibson's second term, and firms pledging their commitment to the city and the mayor were legion (NYT, 12 July 1977, 3 May 1978).[1]

What the Quincy, Newark, and Gary situations have in common is that they involved a loss of control through electoral processes over the formal institutions of government by one ethnic group to another.[2] Furthermore, these transitions of formal power involved political competition between ethnic groups of unequal status in society. The cities in question did not just witness mere changes in personnel, mere transfers of power from one party to another, or simply a change in the social backgrounds of government officials. Rather, they experienced transitions involving ethnic successions in which once dominant ethnic groups, Yankees in Massachusetts and whites in Newark and Gary, whose dominance had hitherto been acknowledged as a virtual birthright, were challenged and effectively displaced in government by ethnic groups that had traditionally occupied a subordinate position in the social and political hierarchy.

How does the once dominant group deal with its loss of formal political power? The manner in which groups, classes, organizations, and individuals generally deal with political defeat—particularly defeat that seems to mark the end of a long period of unquestioned dominance—is a subject that has received scant attention from social scientists. Certainly the phenomenon of losing is common enough in politics, but the perspectives of investigators have been shaped more by questions about how winners handle their accession to power. For those who study American political conflict in a partisan context, a focus on winners is perhaps understandable, not simply because social scientists are interested in how winners consolidate their power and implement their policies, but because the response of losers is relatively predictable. The usual partisan response to the loss of formal power is to regroup and attempt to regain power through electoral contestation.

The kinds of tensions and pulls produced by ethnic conflict, however, suggest a much wider range of possible responses by a defeated

[1]For the key to newspaper names, see the List of Abbreviations.

[2]Milton Gordon (1961) has defined the term *ethnic group* as "any racial, religious, and national origins collectivity [p. 263]." I shall on occasion, however, employ the term *ethnoracial* to assure recognition of its intended inclusiveness.

ethnic group, especially when they may anticipate minority status over the long run. Because ethnic boundaries often divide people along cultural lines, problems of cultural clash are more likely to color ethnic competition and transition than in partisan contests. Because ethnic ties may still elicit very basic loyalties, especially in politics, questions of ethnic pride contribute to the stakes at issue. In addition, since ethnic lines frequently demarcate ancient and deeply rooted animosities, notions of threat and group survival are brought into play in situations of ethnic succession.[3] For all these reasons the vulnerability of the losing group in an ethnic contest is acutely felt and may be expected to give rise to a wider set of strategic responses than conventional electoral contestation. Subversion, withdrawal, cooperation, and maintaining strategies, discussed later in this chapter, may be as attractive and more productive than contestation, although they are not the usual reaction to partisan loss.

The modes of response or adjustment by once dominant ethnoracial groups to the loss of political power to new ethnic majorities and the conditions that shape those adjustments are the principal concerns of this study; specifically, the investigation focuses on the reactions of Yankee elites at the turn of the century in Boston, and on those of white elites in contemporary Atlanta and Detroit. Such an inquiry not only affords historical perspective on the process of ethnoracial political transition in America, but it also covers the two most salient domestic instances of this phenomenon, namely the shift from Yankee to Irish hegemony in the cities of the Northeast, and the developing trend toward black ascendancy in certain contemporary big cities.

The aims of this study may be summarized simply:

1. To provide a descriptive account of the adjustment of displaced elites to the loss of formal political power to previously subordinate ethnic and racial groups
2. To suggest the nature of the implications of the patterns of adjustment for the cities in question as well as for the larger American society
3. To explore the factors that give rise to particular patterns of adjustment to the loss of power, and thus to move toward a theory of ethnoracial political transition

[3]During the mayoral election in Newark in 1970, in which a black man, Kenneth Gibson, was challenging the white incumbent, white Police Director Dominick Spina told a white audience that the election represented "a battle for survival" and argued that "whether we survive or cease to exist depends on what you do on June 16 [NYT, 9 June 1970, p. 30]."

A DEFINITION OF ETHNORACIAL
POLITICAL TRANSITION

The notion of *ethnoracial transition* suggests a process that takes place over time, but the key event that signals a transition is the capture of executive office by a member of a formerly subordinate ethnic group.[4] In urban politics, where political competition among ethnic groups has occurred most visibly and regularly, this means the capture of the mayoralty. Up until that moment the politically emergent ethnic group has generally achieved a variety of pretransition breakthroughs at the level of party office (Cornwell, 1960), in patronage acquisitions (Lowi, 1964), or on the city council (Dahl, 1961). An ethnic group may not, of course, ever manage to gain control of city hall. Only when a working electoral majority develops that can seize and retain the mayoralty can we speak of transition in urban politics.[5]

We may, then, define *ethnoracial political transition* as the acquisition of formal executive office in a political jurisdiction by a member of a previously subordinate ethnic group that is now backed politically by a new, potentially durable working majority composed largely of or dominated by members of that group.

Let us examine the components of this definition. The importance of the capture of the mayoralty is crucial. The chief executive occupies the apex of the government institutional structure and is the critical actor in the urban policy process.[6] The mayor is not only the most visible public official in the city but is in many ways symbolic of its condition. Thus, any mayor—but especially one who is in some way out of the ordinary—represents a potentially arresting, powerful stimulus in the field of an observer's vision. As Murray Edelman (1964) has pointed out:

> Governmental leaders have a tremendous potential for evoking strong emotional response in large populations. When an individual is recognized as a leading official of the state, he becomes a symbol of some or all the aspects of the state: its capacity for benefitting and hurting, for threatening and reassuring [p. 73].

The first achievement of an important mayoralty represents a collective political coming of age for the newly victorious group. Thus, William Grace, the first Irish Catholic mayor in America, elected in New York in 1880, and Hugh O'Brien, the first Irish Catholic mayor of Boston, elected

---

[4] For an elaboration of this point, see Eisinger (1976).

[5] Edward R. Kantowicz (1975), explaining why Chicago Poles had never managed to elect a Polish-American mayor, notes that ethnic politics "only succeeds when the group forms a majority of the voters in a political division [p. 217]."

[6] See the findings of Kuo (1973).

in 1884, performed such a symbolic role for the Irish. Fiorello LaGuardia fulfilled a similar function for Italian Americans with his election in 1933 in New York, as did Hatcher and Carl Stokes for black Americans with their elections in Gary and Cleveland in the late 1960s. But if the accession to power by the first member of a newly emergent group is an energizing and salutary symbol in some quarters, it is regarded with apprehension in others. As Pettigrew (1972) has noted, unlike the situation in which blacks compete for city council positions, a black running for "captain of the ship" poses a "rigorous test of the white voter's racial prejudices and behavior [p. 96]." In the same vein Matthew Holden (1973) has observed that "ethnic newcomers to elective office may be admitted into legislative bodies, even though their entry into executive positions would be resisted. [Executive positions] permit their holders to acquire bargaining parity, rather than clientship, because these holders are able to exert some control over the claims which other people would wish to make [p. 105]."

The definition of *ethnoracial political transition* thus stresses the capture of the mayoralty as an important event, but one may not thereby presume that all of the old patterns of influence and power have inevitably been altered. Ethnoracial political transition focuses on the transfer of formal authority as the result of the emergence of a new electoral majority. Patterns of dominance and subordination based only on the composition of the voting majority and the characteristics of the holders of office can be presumed to have changed, but there is a possibility that those who no longer control the formal apparatus of government will still manage somehow to exercise dominance through means other than city hall. Nevertheless, it is probable that the loss of formal authority and the descent to minority status in a situation of ethnoracial transition will bring about a significant loss of power and influence on the part of the group formerly in possession of the mayoralty and its associated perquisites.

If the impact of ethnoracial transition on the structure of power is an open question, however, such changes almost inevitably occasion new political adjustments, the development of new perspectives, and the search for new or compensatory strategies of influence by those who have been displaced. Ethnoracial transition suggests a fluid situation, in which traditional relations among groups and habitual modes of acting in politics undergo transformation.

Two other elements of the definition of *ethnoracial political transition* require clarification: durability and working majority. Transition cannot be said to have taken place unless there is a strong likelihood that the new group will maintain its new dominance over some period of time. Naturally, the precise number of years that distinguishes durable from temporary dominance is not ascertainable. Common sense notions suggest that in urban politics durability involves control of city hall by a particular ethnic group for more than a couple of mayoral terms, and that that ethnic group can capture the mayoralty with a candidate different from the initial pioneer victor.

In Cleveland black rule was temporary rather than durable, at least in the short run. Carl Stokes won two 2-year terms, beginning in 1967, in a city where the black electorate amounted to slightly less than 40%. In 1971, however, the city reverted to white rule when Stokes declined to run again. Black government in Los Angeles can also be seen as a temporary phenomenon. Despite the apparent popularity of Mayor Thomas Bradley (Roberts, 1974), the black population, which amounts to little more than one-fifth of the city's population, could scarcely hope to maintain a black in city hall after Bradley.

Durability is clearly related to the existence of a working majority, which is a function of either numerical or organizational dominance. The Chicago and the Boston Irish offer contrasting illustrations. The latter achieved numerical superiority around the turn of the century and maintained it for at least 60 years. The Chicago Irish, however, who enjoyed only a brief period of majority status in the nineteenth century, were nevertheless able to establish durable control through their organization and dominance of a party machine whose working majority included other groups besides the Irish.

It is too soon to assert the durability of black rule in places like Detroit, Atlanta, Newark, and Gary, but a major condition for long-term black government—a black majority—is present in each of these cities. Given the high probability of continued black rule where there is a black majority, it is reasonable to argue that an ethnoracial political transition that meets the minimum requirements of our definition has occurred in such places.

THE IMPORTANCE OF STUDYING THE DISPLACED

Groups displaced from positions of formal political power do not ordinarily vanish from the scene, at least in the period immediately following transition. This is especially true in American cities, where displaced elites have not been subjected to official programs of harassment or repression. It is possible that transition may accelerate ongoing patterns of out-migration from the city by mass and elite members of the formerly dominant group. But whether or not this is the case, the changes that are occurring in the demography of contemporary cities as a result of migration patterns appear to be taking place slowly enough to allow formerly dominant groups to remain almost as large as the newly dominant group in the initial years of the transition.

The continued presence in the city of large numbers of members of the formerly dominant group, especially people who still occupy or have recently occupied elite positions in politics, finance, business, and the professions, makes this an important group to study for three major reasons:

1. The response of the formerly dominant group to displacement bears on the ability of the newly dominant group to govern effectively
2. The response of the displaced bears on the economic, cultural, and psychological future of the city
3. The response of the displaced also casts light on the larger context of ethnoracial conflict and cooperation in America

IMPLICATIONS FOR THE NEW GOVERNORS

Any group accustomed to ruling will possess various important skills and resources, or at the very least, ways of commanding them. Certainly this is the case with regard to the white communities of modern Atlanta and Detroit, and the Yankee community of Boston at the turn of the century. The thorough dominance of members of these groups in commerce, banking, industry, real estate, law, and the press persisted in Boston, and still persists in Detroit and Atlanta, through the ethnoracial transition in government. Prior to transition, these economic and professional elites either provided the recruitment pool from which local government elites were drawn, or at least tended to share with politicians certain basic perspectives on the city and its operations. Under these circumstances government elites in the three cities could count upon tapping certain of the administrative, intellectual, and financial resources, as well as the prestige and influence of the economic and professional sectors as they sought to govern. For lobbying trips to the state and national capitals, the appointment of panels and commissions, the launching of development projects, the recruitment of high-level bureaucrats from the outside, the attraction of investments, conventions, and business to the city to enhance employment opportunities and the tax base, and so on, the politicians could often count on the availability, aid, support, initiatives, and ready advice of the nongovernmental elite sector.

This does not mean that common ethnicity or race eliminated conflicts in the government, economic, and professional communities, or produced a thoroughgoing homogeneity in the civic culture. All that it suggests is that so long as government and private sector elites shared ethnic backgrounds or race, at least there were no ethnic barriers to hinder the flow of resources from one group to the other. Ethnic considerations did not enter into calculations of the accessibility to political leaders of resources controlled by elites outside of government. The key question is whether these resources are still available to government officials in the period after ethnoracial transition has occurred.

As the first blacks began to win important mayoralties, general expectations were that the new governors would not have ready access to such resources. Charles H. Levine (1974, p. 37) suggested in an early study of black-mayor cities that white hostility had already constrained those black mayors who sought to exert community-wide leadership, resulting in a kind of political immobilism. A month before the breakthrough mayoral elections in Gary and Cleveland in 1967, Frances Fox Piven and Richard Cloward (1967) predicted that in the event of black control of big cities, "Millions of whites unable or unwilling to leave will remain in the core cities, a fact of key political importance, since they will fiercely resist the exploitation of municipal power for black interests [p. 17]."[7]

Resistance to black rule by those whites who remained in the cities was not the only way that observers expected the displaced group to deny its resources to the newly victorious black politicians. Retreat from the city to the suburban ring, involving the actual physical withdrawal of resources, was seen as another likely development (NYT, 18 Oct. 1973). In addition, there appeared to be at least two other means by which it was anticipated that displaced whites would constrain black rule. One was through the denial of state aid by white state legislatures to black-run cities. "As Negroes assume positions of political power within municipal governments," Paul Friesema wrote in 1969, "it seems altogether probable that state legislatures, mostly whites representing other whites, will become even less interested in providing funds or

[7]The observation was not a novel one. A decade earlier Morton Grodzins (1958) had written, "In the long run, it is highly unlikely that the white population will allow Negroes to become dominant in the cities without resistance [p. 14]."

other aids to cities [p. 77]."[8] The constraining impact on black leaders would be immediate. Fearing to offend the legislatures upon whom they are so dependent, black political leaders would have to temper or curtail the very militance that may have made them so attractive to black constituencies in the first place (p. 77).

The other predicted strategy for constraining black political power involves the denial to black central city politicians of control over a variety of fiscal, legal, and political resources through a push for metropolitan government (Piven and Cloward, 1968, p. 23). Metropolitan reform in the rationalistic tradition normally involves either the superimposition of new metropolitan authorities (e.g., single or multipurpose special districts) or the enlargement of the territorial jurisdiction of existing central city government machinery. In either case important elements of local government become accountable to a new majority white constituency. The effect in black majority central cities would presumably be to deny black leadership the majorities needed to stay in power, as well as to strip from it control over governmental functions and authority, newly lodged in the hands of recently created metropolitan bodies. Whites

[8]The evidence that state legislatures would pursue such a course is mixed. In the following table state revenue as a percentage of each city's total revenue is averaged for the 4 years prior to the election of a black mayor, and separately for the period of black rule. As a crude control, the same averages are calculated for the same time spans for another city of comparable size in each state, but which had unbroken white rule. The case of Cleveland seems to conform, marginally, to the hypothesis advanced by Friesema, although the control city in Ohio shows inexplicable patterns in the context of this argument. Gary and Newark do not conform, nor do data presented in Chapter 5 on Atlanta and Detroit support the hypothesis. The case for rejecting Friesema's hypothesis is stronger than that for accepting it.

State Revenue as an Average Percentage of
Total Municipal Revenue

	Ohio		New Jersey		Indiana	
Time period	Cleveland[a]	Columbus	Newark[a]	Elizabeth	Gary[a]	South Bend
Four years before election of black mayor	8.4	14.0	21.0	16.7	14.5	13.3
Period of black rule	5.4	13.3	42.3	21.9	14.8	16.2
After black mayor[b]	10.9	10.5	—	—	—	—

Source: City Government Finances (1964–1976).
[a] Black-mayor cities.
[b] Black rule ended in Cleveland in 1971. It still continues in Newark and Gary.

displaced from positions of formal political power in the central city could quite reasonably justify the diversion of resources they control from the black governors of the central city core to those leaders who operate at the higher metropolitan level.

In short, the displaced group still constitutes a force in the city with which to be reckoned. By denying, removing, or diverting its resources, it may preclude effective government by the new victors, for the resources controlled by that group alone are likely to be inadequate by any standard. An examination of the disposition of the displaced group's resources assumes even more critical importance in the present cases, for the social class disparities between the new black victors and their constitutents on the one hand and the displaced whites on the other are so acute. Such an investigation, then, throws light not simply on the dynamics of displacement in ethnoracial political contests but also on whether the newly empowered representatives of traditionally impoverished groups are left to generate their own meager resources or whether the resources—or some of them—controlled by members of the displaced group are available for use.

IMPLICATIONS FOR THE CITY

Few, if any, considered the possibility that the advent of black mayors in big cities might hasten a return to social peace after the turbulence of the 1960s. It would have been perfectly plausible to predict that black victories in the conventional arenas of electoral politics would bring substantial relief to beleaguered white communities after the tense years of confrontation by assuaging black disaffection, wedding blacks to the system through which they had finally gained, and dissipating radical energies in the black community. Instead, however, the transition to black rule was anticipated with fear and apprehension. "How much of a disaster would it be if some of our major cities were to become predominantly Negro?" Irving Kristol asked in 1966; "I am sure we would find the prospect disturbing."

Much of the concern over the impact of black mayoral victories focused on the anticipated effects on race relations and thus on the morale and social climate of the cities in question. Piven and Cloward (1967) were convinced that racial polarization would sharpen: "Negro control can only deepen racial cleavages in the urban area . . . . Black majorities also mean the alienation of urban whites [p. 17]." On the morning after Maynard Jackson's first mayoral victory in Atlanta in 1973, *The New York Times* (18 Oct. 1973) wondered editorially whether whites, fleeing to the suburbs with their wealth and civic experience, would "abandon the city to a fate that perhaps only General Sherman might have wished for it."

Voting patterns in cities where black candidates defeated whites for control of city hall revealed, predictably, a major gulf between voters of the two races. Although survey evidence from the mid-1960s had suggested that a majority of white voters would be willing to cast their ballots for a "capable" black running for mayor, when it came to the test in Gary, Cleveland, Newark, Atlanta, and Detroit, no more than an estimated high of 22% and as little as 8% of the white electorate selected the black candidate.[9] At the same time more than 90% of black voters were choosing the black candidate. In partisan elections in Gary and Cleveland, Democratic whites deserted their party's black nominee to vote for his white Republican opponent (Pettigrew, 1972, pp. 99–100). Not only was racial bloc voting the common pattern in these contests, but voters also turned out in unusually high numbers. By all indications, then, it would appear that the critical event in ethnoracial transition in these modern cities occurred under conditions of high racial polarization, at least at the mass level.

There is no question that racial polarization can demoralize a city and set its inhabitants on edge, but it is possible to ameliorate these tensions and minimize their practical consequences.[10] Whether or to what extent polarization will affect a city's economic, cultural, and spiritual life depends greatly on its civic leadership. The burden is particularly heavy for a recently displaced elite, which can either bridge or exacerbate the racial gap by its actions and rhetoric. Members of the displaced group may withdraw from politics and civic affairs, fight the new governors in a variety of ways, or cooperate with those who control city hall. They can affect the economy and tax base of the city through their decisions to move, stay, expand their business enterprises or invest their money. They may embrace a rhetoric of confidence or of demoralization, thus affecting the image of the city as a site for investments from the outside, conventions, tourism, and trade. To the degree that members of the displaced group maintain a genuine commitment to the city in which they live and work, the deleterious consequences of racial polari-

[9]For surveys of whites' willingness to vote for blacks, see Campbell and Schuman (1969, p. 34), and Devine (1972, p. 337); for actual voting patterns in mayoral elections involving black candidates, see Pettigrew (1972) and Masotti et al. (1969); see also NYT (17 Oct. and 8 Nov. 1973), DN (7 Nov. 1973), and AC (17 Oct. 1973).

[10]The contrast between the white elites of contemporary Boston and Detroit as both cities approached busing of school children is instructive. In Boston elites sought to capitalize on white resentment, encouraging demonstrations and defiance. In Detroit a year-long public relations campaign to obey the busing order that took effect in early 1976, orchestrated by a broad coalition of elites, contributed to its smooth implementation.

zation can be minimized and the process of ethnoracial political transition can be smoothed. Much of the quality of life in such cities depends, then, on the grace of the losers.

IMPLICATIONS FOR UNDERSTANDING THE CONTEXT OF ETHNORACIAL POLITICAL CONFLICT

A strong basis for judging the character and moral basis of a multiethnic society, a category that includes all but a dozen or so modern nations (Connor, 1973), is the way in which the various ethnic groups deal with one another in competitive situations. Discrimination, genocide, unofficial but sanctioned violence, and apartheid are common strategies for managing ethnic competition in many places in the world. Societies in which ethnic conflicts are resolved in a peaceful, equitable, and nonvindictive way are comparatively few in number. Historically, at least, the United States could not have been included among them.

Given the tradition of interracial and interethnic violence in America, however, it is striking that ethnoracial political transition—both in Irish Boston at the turn of the century and in the modern cities that concern us—was accomplished peacefully, a matter to all appearances of electoral routine. Undoubtedly, the American faith in the sanctity of election outcomes has something to do with this peaceful acceptance of the victories of much feared and often scorned groups; nevertheless, it is worth exploring not only the extent to which this has been a governing factor, but also the deeper forces that may have held displaced elites in particular to this faith.

That political succession can be accomplished peacefully in a culture where passions in racial conflicts run high is remarkable in itself and warrants investigation. It is also important, however, to explore the dynamics of the subsequent adjustment of the displaced as a way of gaining insight to the accommodative capacities of the society. During the course of American history it has not been uncommon to find that the dominant groups responded to potential ethnic challenges in the political and economic spheres with violence (Eisinger, 1976). This sordid pattern suggested that American society might not be the permeable political and economic opportunity structure that an egalitarian ideology required. Certainly the pattern held as blacks began to make assertive strides during the middle years of this century.

Once an ethnic group has achieved a certain degree of entry into the system, however, violence against it by the dominant group has tended to diminish, eventually disappearing altogether as a means of social control (Eisinger, 1976). Modes of competition in politics become conventional, nonviolent, and relatively predictable, at least in cases involving white ethnic groups. Electoral contestation between newly powerful ethnic groups and their antagonists becomes the reigning norm. Displaced groups without much hope of regaining power or once dominant groups now forced to share power seem to have sought not so much to subvert the new order as to find a place in it. The question in the present case is whether such accommodation will hold in a situation of racial transition. If so, then we might conclude at least that American society is ultimately more pliant and open than the convulsions that herald incipient changes in the ethnic or racial balance of power would suggest.

It is also possible that an examination of the response of the displaced may supply clues as to the contemporary nature of the American commitment to the survival of the big cities. Americans have never loved cities, nor have they, historically, expended much unusual effort to keep them healthy (White and White, 1962). Most people, including those who live in cities, would prefer to live elsewhere.[11] One might suppose, then, that people would seize upon the least excuse to desert the central city. It would not be unreasonable to predict, given the historic context of racial animosity and the generalized hostility toward urbanism as a way of life, that the capture of political control by blacks would provide such an excuse, making for wholesale white withdrawal from the life of the core. In such a situation concerted reconcentration of resources and energies on the metropolitan fringe would represent a decision to abandon the city, perhaps to divest it of its ancient cultural and economic functions. The response of the displaced to their new status, then, offers evidence on the extent to which the powerful in white America are willing to let the cities slide unsupported toward an uncertain and penurious future. In short, a study of the displaced may provide important and perhaps fresh perspectives on some of the larger outlines of American society, including both its accommodative capacities and its commitment to urban life.

EXPLORING THE PROCESS OF ETHNORACIAL TRANSITION

In the extensive literature on the politics of race and ethnicity, the phenomenon of transition has scarcely been investigated from the point of view of the displaced group. In the theoretical writings on ethnic relations, several scholars recognize explicitly that the statuses of vari-

[11]See data cited in Dahl (1967, p. 965) and in a recent Gallup poll reported in NYT (2 March 1978).

ous groups may not be fixed in the stratification system. Shibutani and Kwan (1965) note that "in some cases ethnic groups change their relative ranks [p. 351]," and William Petersen (1975) has written, "Nothing is so likely to exacerbate interethnic antagonism as what the French Canadians call *la revanche des berceaux*—the vengeance of the cradles: with a differential growth rate every settlement is tentative, in force only till the day—eagerly awaited or fearfully dreaded—when the minority and the majority change places [p. 195]." Enloe (1973) has observed that a dominant group may not "operate in a static society, in which status distinctions between ethnic communities are fixed [p. 209]." But this recognition leads Enloe at least to suggest that the problem inherent in such a state of social flux is how the dominant group faces a challenge to its status from below, although ignoring the possibility that that group might actually have to adjust to the loss of power. "Ruling communities," she writes, "must adjust to the rising political awareness and expectations of the ethnic groups they traditionally dictated to or ignored [p. 209]."

Scholars cannot explain their neglect of ethnoracial political transition by arguing that such occurrences are rare. Virtually every case involving the emergence of new nations from colonial status in the Third World provides an analogue of sorts to the ethnoracial transitions we have identified in American cities. The French in North and West Africa, the Dutch in the East Indies, the Belgians in Central Africa, and the British in East Africa all represent once dominant groups displaced from power by the emergence of majoritarian rule. Although it is true that the colonial system was not overturned by electoral processes, colonial rulers were nevertheless politically displaced. How they responded to the loss of dominance is of critical importance, as I have argued. There is some evidence that colonial elites responded with concern in anticipation of the loss of power: Tanner (1966), for example, noted the deterioration of morale among Europeans as independence approached in Tanganyika. But for the most part students of decolonization and the adjustment to independence have not been interested in the psychological or strategic responses of the colonial group once it has been displaced (see, e.g., Zolberg, 1964; Young, 1965; Apter, 1963).

One notable exception is Donald Rothchild's *Racial Bargaining in Independent Kenya* (1973), a book that focuses on race relations during and after the change from European to African domination. Prior to independence in 1960, Kenya, like other African colonies (as well as many American cities before the emergence of black and Irish power), was a stratified society in which the structure of privilege largely coincided with ethnic and racial distinctions. Once out of power after independence, the Europeans and Asians were beset by a host of unfamiliar anxieties as they were forced into a quest for status, influence, and security in a situation where the rules had suddenly changed.

Rothchild is interested both in the modes of adjustment of the once dominant groups to the loss of power, and the implications of their adjustment for the newly independent nation. The psychological adjustment of the Asians and Europeans was influenced by the recognition of their newly marginal status. Hence their early responses were to worry about gaining citizenship in the new nation, about discrimination in the application of immigration and trade licensing laws, and about the potential denial of business and employment opportunities (Chapters 10–11). In what is perhaps an ironic turnabout, Rothchild points out, many non-Africans were led to a commitment to "symbols of legal equity to castigate any evidence of racial discrimination between citizens [p. 365]." A major strategic response for many was to leave the country altogether. Until independence, permanent European immigration exceeded emigration, but after 1960 the balance shifted so that two Europeans left for every one that entered (Chapter 12). Many non-Africans stayed in Kenya, however, and lent their skills and capital to the new regime.

In the end, ethnoracial political transition in Kenya was accomplished without a breakdown in the social order. "Obviously, men recognized a greater mutuality of interests than was previously anticipated; the decline in minority visibility and the increase in African opportunity acted as tension relievers; the fears of all groups were to some extent unrealistic [p. 145]." But this in no way diminishes the importance of those fears. What Rothchild's study teaches us is that the salience of such concerns among the displaced was great enough to change some very basic perspectives on the nature of the society in which they were living and on their own roles in it. Furthermore, the Kenya case suggests the importance of investigating a society's accommodative capacities in the face of ethnoracial transition, and indicates the necessity to search for the conditions that put those capacities, such as they might be, into play.

Studies of ethnic politics in America offer as few guidelines for the exploration of transition and displacement as the literature on decolonization. Characteristically, such studies lack breadth, theory, and detail pertaining to the experience of displacement in particular. Nevertheless, the field is not entirely barren; a number of people have at least recognized some of the dilemmas of the displaced. Pratt (1970), for example, has examined the Protestant Council of the City of New York as it became an organizational vehicle for white Protestants, "who once enjoyed deference and special status as a result of their near-monopoly of

'American' traits [p. 223]," dealing with the increasing political threat in the postwar period of second- and third-generation immigrants.[12]

Kenneth Underwood's more extensive study (1957) of Protestant–Catholic relations in Holyoke, Massachusetts, is motivated by similar interests. Underwood is concerned with what happens to an American community "when it becomes Catholic [p. xv]," that is, when Protestants are placed in an unaccustomed minority position, and notes that in the New England setting especially, these religious divisions largely coincided with the classic Yankee–Irish ethnic cleavage (pp. 207–208). The investigation focuses primarily, however, on the theological and church organizational dilemmas that confront the Protestants as they struggle with an almost obsessive absorption with Catholic dominance (p. 369): They worry about religious survival, show an "unusual concern for religious diversity and civil liberty," and exhibit a good deal of uncertainty about their religious mission (pp. 371–374).

The theme of Protestant Yankees under siege and in decline is most fully touched upon in the literature on Boston culture and politics at the turn of the century. These works, primarily historical, provide a starting point for Chapter 2 and are cited in more detail there. Taken individually, it is important to point out, they offer only a fragmentary and largely atheoretical perspective on the Yankee adjustment to displacement. Blodgett (1966), for example, is mainly interested in Yankee Democrats during the transition to Irish dominance in Boston during the 1890s, whereas Abrams (1964) is concerned with Yankee Republicans in the next decade. Barbara Solomon (1956) provides an account of the intellectual underpinnings and formation of the Boston Immigration Restriction League, Arthur Mann (1954) writes of Yankee reformers, and Martin Green (1966) discusses the cultural dilemma that confronted Brahmin intellectuals in a radically changing society.

Hofstadter's account (1963, pp. 172–179) of the alienation of New England gentlemen intellectuals, their fall from positions of power, and their brief ascendancy (again through reform) plays on the same theme of Yankee displacement. Hofstadter views this group, however, not as participants in an ethnic struggle but rather as a class displaced by a new plutocracy and the professional politician (pp. 176, 408). Dahl (1961) recounts the same story in his study of New Haven, though with a more certain sense of the ethnic struggle involved.

[12]Another well-known study of Protestants on the defensive is Digby Baltzell's *The Protestant Establishment* (1964). Baltzell treats Protestants not so much as a displaced group, however, but as a group threatened by American ethnic heterogeneity. He focuses only on one major defensive response to this threat, namely, the retreat to a caste society built on the invidious base of anti-Semitism.

In none of these studies have scholars sought to explore a wide range of adaptations to displacement, the patterns of variation, or the conditions of adjustment. They provide, then, a backdrop for the present inquiry. They reassure us of the salience of the problem of ethnoracial transition in American life, and they record for us, as we explore more contemporary examples of this phenomenon, the continuities in the process.

SOME THEORETICAL POINTS OF DEPARTURE

In surveying the fragments that bear on ethnoracial transition from the point of view of the displaced, I have not speculated on why scholars have not explored the process from this perspective more fully. To answer this question, we must glance briefly at the theoretical literature on race and ethnic relations, which will also provide a clearer view of the departures made in the present study. Three broad problems have characterized the dominant thinking about ethnic relations and have probably served to divert investigators from a focus on the displaced: These might be called The Bandwagon Effect, The Liberal Expectancy, and A Failure of the Imagination.

The Bandwagon Effect refers to the magnetic political—or in this case, intellectual—excitement that winners generate. The first moment at which we might have expected modern social scientists to become aware of displacement in ethnoracial political transition was during the late 1950s as decolonization in Africa got under way. But, quite understandably, the "beacon of independence," as Zolberg has called it, blinded students of the new nations to questions concerning the adjustment of the old order. What was interesting, attractive, and ideologically more comfortable to contemplate was the new order. As Zolberg (1966) has written:

> Most political scientists who were in the field sufficiently early to share in the enthusiasm of the new men at the helm of liberating movements . . . were caught up in the drama of man's search for polity which was being reenacted in a new and strange environment. The study of African politics provided a great and exciting intellectual adventure comparable to the quests which earlier had driven explorers to overcome apparently insurmountable obstacles on the same continent [p. 1].

It is true that white colonials still possessed vast economic and skill resources in the new nations, although they comprised a small minority.

But the academic preoccupation with nation-building simply stripped these small white communities of intellectual significance. To have wondered about them in any serious way, even to ask what it was that they might have contributed to the new regime, would have been, in the climate of which Zolberg speaks, to appear to pursue a very minor and not entirely acceptable tributary in the watershed of scholarship.

The Liberal Expectancy, Milton Gordon's term (1975, p. 88), is an entirely different matter. It involves the expectation that ethnic and other ascriptive distinctions among people would eventually give way to achievement-based stratification and national cultural homogenization. Gordon describes it as the familiar sociological assumption that the primordial ties that divide people will be extinguished under the impacts of urbanization and modernization. Intergroup contacts, in this view, would ultimately lead to assimilation. One early exponent of this argument, Robert Park (1950), asserted that "the race relations cycle, which takes the form . . . of contacts, competition, accommodation and eventual assimilation, is apparently progressive and irreversible [p. 55]."[13] Assimilation, by which Park meant both acculturation and the structural incorporation of out-groups into the larger society, would be the consequence of personal intercourse and friendships that would undermine barriers of racial segregation and caste (pp. 150, 204, 209).

By a somewhat different route, Robert Dahl (1961, pp. 35–36) did much to perpetuate the supposition that assimilation lay at the end of the course of ethnic contacts in American society. His argument was that ethnic identity, as a force in shaping the political behavior of any given group, would survive only until a significant segment of that group acquired middle-class status. At this point class perspectives would not only lead to a gradual rejection of ethnic politics as embarrassing and irrelevant, but they would also help to bond the upwardly mobile ethnics to others in society who had risen in the class structure.

It is true that for some ethnic groups in America, as well as in other places, assimilation has occurred. But for the most part the color-blind society and the melting pot are chimeras—unreal, and what is more, unwanted. As Philip Mason (1970) points out in regard to the assimilation model, "The concept assumes smooth unbroken progress towards harmony, but today the world-wide revolt against hierarchy and fixed status is everywhere producing movements in the reverse direction [p. 55]."

The more one expects assimilation to occur over the course of ethnic

[13]As Philip Mason (1970, p. 55) points out, this is not, properly speaking, a cycle.

relations, the less likely it is that one will seek to develop alternative models or search for contradictory evidence.[14] Nevertheless, some students of ethnic relations were never wedded to the assimilation argument, but used it only as a foil in a hunt for alternatives. Here it can only be suggested that *A Failure of Imagination* occurred, for none of the extant alternative projections of the outcomes of ethnic relationships suggested the possibility of what we have called *transition*, that is, a turnabout in the formal patterns of political domination and subordination.

The most common alternatives to the assimilation argument are conflict models. Intergroup contact is thought to lead not to assimilation but to competition or conflict in a pluralistic society. Milton Gordon (1964) suggests, for example, that although massive acculturation has occurred in America, structural pluralism resting on an ethnic base remains. Similarly, general conflict theories, such as those of Michael Banton (1967) or Pierre van den Berghe (1967), posit a movement or transformation in ethnic relations from contact to some form of domination to an end-state at which competition and pluralism prevail. For van den Berghe, the formerly dominant group is seen as maintaining some degree of superiority and advantage, even as it competes with those groups it once thoroughly dominated (pp. 99, 127). Banton, however, argues the possibility of true pluralism as one outcome of intergroup contact, where race is no longer the basis of invidious distinction and no group is a dominant competitor (pp. 73–74). What is important about these models is that the potential transformation of a once dominant group to subordinate status is never considered.[15]

To summarize, I suggest that what emerges in the theoretical literature are three very basic models of ethnoracial relations: *traditional hegemony*, *pluralism*, and *assimilation* (see Figure 1.1; the three are not assumed to represent a developmental sequence in this presentation). Our concern in this present study, however, is with a fourth model, which may be called *succession* (Figure 1.1). Although succession, as well as the other models of ethnic relations, encompasses a broader range of relationships than those related simply to the acquisition of control over the formal apparatus of government, our specific concern is political succession.

Each of these four models suggests different research foci for the

---

[14]Pierre van den Berghe (1967) writes that the field of race relations "has been dominated by a functionalist view of society and a definition of the race problem as one of integration and assimilation of minorities into a mainstream of a consensus-based society [p. 7]" (cf. Schermerhorn, 1970, p. 14).

[15]The possibility of this is mentioned in passing by Blalock (1967, p. 189), but it is never developed.

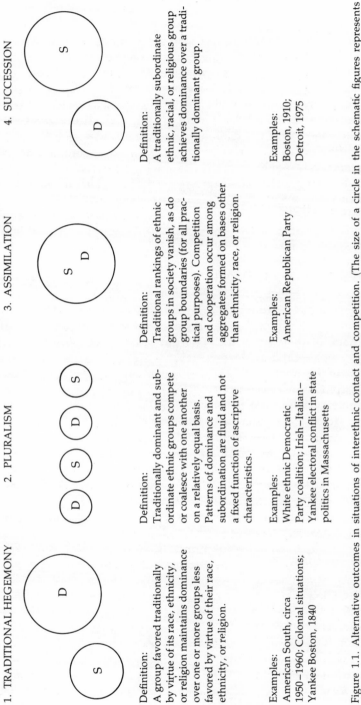

1. TRADITIONAL HEGEMONY

Definition:
A group favored traditionally by virtue of its race, ethnicity, or religion maintains dominance over one or more groups less favored by virtue of their race, ethnicity, or religion.

Examples:
American South, circa 1950–1960; Colonial situations; Yankee Boston, 1840

2. PLURALISM

Definition:
Traditionally dominant and subordinate ethnic groups compete or coalesce with one another on a relatively equal basis. Patterns of dominance and subordination are fluid and not a fixed function of ascriptive characteristics.

Examples:
White ethnic Democratic Party coalition; Irish–Italian–Yankee electoral conflict in state politics in Massachusetts

3. ASSIMILATION

Definition:
Traditional rankings of ethnic groups in society vanish, as do group boundaries (for all practical purposes). Competition and cooperation occur among aggregates formed on bases other than ethnicity, race, or religion.

Examples:
American Republican Party

4. SUCCESSION

Definition:
A traditionally subordinate ethnic, racial, or religious group achieves dominance over a traditionally dominant group.

Examples:
Boston, 1910;
Detroit, 1975

Figure 1.1. Alternative outcomes in situations of interethnic contact and competition. (The size of a circle in the schematic figures represents relative power, not size of group. D = traditionally dominant group; S = traditionally subordinate group.)

political scientist. In examining a situation of traditional hegemony, for example, one is drawn to investigate the means by which the traditionally dominant group maintains its dominance (e.g., racism), or the devices that the traditionally subordinate group employs to change or resist its status (e.g., protest). A pluralistic system suggests a focus on strategies of coalition-building and the dynamics of group conflict in a society of relative equals. An investigation of an assimilated society leads one to explore the devices and pressures that lead to and maintain integration. Two general research foci emerge in studying succession: the character of the traditionally subordinate group's adjustment to its new political dominance (on which much has been written) and the nature of the traditionally dominant group's adjustment to its new subordinate status (which is the concern in this analysis).

MODES OF ADJUSTMENT

To speak of adjustment in our particular urban context is to focus on the ways in which members of a displaced group come to terms with the new shape of their city's sociopolitical order. Such adjustments have both psychological and strategic (that is, action-oriented) dimensions. The task of this analysis is to describe the patterns of these adjustments and to explore the conditions that determine them.

One way to explore the psychology of adjustment among the displaced is to focus on their perceptions of transition and its most visible symbol, the new mayor. Ethnoracial political transition may be understood as a stimulus for rethinking the nature and significance of one's role in the society of the city. It is simply no longer possible for whites in black-mayor cities, particularly those whites in elite positions, to assume that they bear the same relationship to the structure of authority as they did prior to the emergence of black political control. The key in this rethinking process is the degree to which transition is seen as threatening to the individual and his or her interests. How transition is regarded on a threat continuum is likely to determine the character of one's psychological adjustment.

Transition may not, of course, be regarded as a threatening occurrence; an individual may not feel that black rule promises deleterious consequences of any sort. This does not mean, however, that the individual's role in the social structure does not change, and change requires adjustment. To the extent that transition is perceived as nonthreatening, this adjustment is not likely to be problematic, most probably conforming to one of two patterns—*satisfaction* or *acceptance*. In the former state an individual is gratified in some sense by the fact of transition. Black control may represent for such a person the realization of certain social justice or equity values (e.g., "majorities should rule," or "it's the turn of blacks now"), or it may be seen as a stimulus for the easing of dangerous racial discontent. Thus, *satisfaction* may reflect a particular kind of psychological congruence, where reality has become consistent with or has fulfilled certain beliefs about what ought to be, or it may reflect simple relief that a tense condition has been ameliorated. In contrast, *acceptance* implies less positive affect than *satisfaction* does. It suggests acquiescence to a situation, resignation to what must be rather than satisfaction at what ought to be. It does not, however, imply bitterness, but rather equanimity.

For some members of the displaced group transition is likely to be threatening to some degree; therefore, psychological adjustment takes other forms. *Rationalization* is a form of adjustment in which the threat that is felt is suppressed. The rationalizer argues that nothing in his or her world has changed in any meaningful way. Since certain important aspects of the world have quite visibly changed with ethnoracial transition, however, to maintain that all is entirely as it was is to hold to a false reality.

When the sense of threat an individual may feel is not suppressed, psychological adjustment is likely to take the form of *fear* or *rejection*. The fear adjustment is dominated by a concern for the survival of a preferred life-style and the maintenance of a particular value structure; it is a response to unwanted changes in a cherished pattern of life. Rejection, on the other hand, is dominated by revulsion. It rests in a racist or nativist matrix that assumes the superiority and inferiority of different races and ethnic groups, and the incompatibility of groups on the opposite sides of these divides.

Psychological adjustment to transition is accompanied by *strategic adjustment*, which is broadly defined as planning or taking action in regard to the changed circumstances to protect or promote one's interests. There are five general types of strategic adjustment, each of which has several distinct variants: *cooperation, maintenance, contestation, withdrawal,* and *subversion*.

Cooperation involves a decision to work with the newly dominant group, either in a partnership or as a follower, in which the role of the displaced is determined by formal or tacit agreement. Participation by whites in a black-mayor city in electoral coalitions, on urban task forces, or as friendly advisors to city officials are examples of cooperation. What is important in such an adjustment is that the legitimacy of the new group's dominance is acknowledged. Thus members of the displaced group who choose to cooperate assent to the political leadership of the new group and treat its initiatives as legitimate and preeminent definitions of the agenda for public action.

A *maintaining adjustment* refers to actions designed to preserve certain behavior patterns, relationships, and prerogatives that existed prior to transition, without necessarily seeking consciously to affect the power of the newly dominant group. In fact, the effect of such adjustments is often to enhance the new group's power, but this is not the chief aim of maintaining strategies. A maintaining strategy is an attempt to establish and maintain diplomatic relations between two groups who may be inclined to mutual suspicion but who can profit from a relationship. Major examples of maintaining strategies are some of the fund-raising activities by white elites on behalf of black mayoral candidates, as well as efforts to maintain access to the mayor.

Contestation refers to efforts to engage in open conflict or opposition to the new rulers. Differences in priorities and policies are publicly voiced. Attempts by the displaced group to reassert dominance or at least to have its preferences prevail are openly pursued. Electoral challenge or the search for a challenger, where the candidate is a member of the displaced group, is one form of contestation. Another form may be designated as loyal opposition, in which the displaced group assumes a critical stance in regard to the new rulers and promotes its views, but does not seek to subvert or overturn the new structure of authority. The news media become important in this variant of contestation, for they may be used as a voice for opposition.

A fourth form of adjustment, *withdrawal*, involves some form of "exit" (Hirschman, 1970), either physical removal or the abandonment of customary patterns of civic involvement. In either case, withdrawal results in diminishing the pool of potential resources that the new rulers may call upon. Moving one's business enterprise out of the city is an obvious form of withdrawal, as is changing one's residence beyond the city limits when that change results in a reconcentration of energies and interests elsewhere. Another form of withdrawal is removing oneself from accustomed civic activity to turn one's energies to private rather than public concerns.

The employment by displaced actors of covert or oblique means to reassert formal dominance or to develop or maintain a hegemonic power base outside of government is called *subversion*. There are several major variants of subversion. One involves the search by members of the displaced group for an "acceptable" candidate from the newly dominant group—one who is predominantly dependent on the displaced and controllable by them—to pose an electoral challenge to those leaders or candidates selected by the new group itself. Subversion is attempted here through cooptation. This subversive strategy involves a certain degree of risk, for it requires an acknowledgement of the changed electoral balance in the city, and banks on the ability to manipulate the new majority for the ends of the displaced. Another subversive strategy is the attempt to consolidate significant control, especially over the city's economic fortunes, in some exclusive extragovernmental body or clique. Certain business groups in particular, both formally and informally organized, offer such potential bases of control. To the extent that this robs local government of influence over the city's revenue-raising capacity and economic development, the power of the new group may be greatly constrained. A third variant of the subversive adjustment is to seek to change the powers or structure of government in a way disadvantageous to the new majority. Lodging formerly municipal powers in the hands of state government or pursuing metropolitan reform are subversive in this sense.

Table 1.1 summarizes the different types of psychological and strategic adjustments. Once the dominant patterns in a city have been ascertained, we should be able to make statements about the degree to which adjustment patterns are monolithic or pluralistic and why, the degree to which resources controlled by the displaced group are available to the new rulers, the implications of transition for the morale and life of the city, and the nature of the accommodative capacities of American society as they apply to cities undergoing ethnoracial political transition.

TABLE 1.1
Types of Adjustment to Ethnoracial Political Transition

Psychological adjustments	Strategic adjustments
Satisfaction	Cooperation
Acceptance	Maintenance
Rationalization	Contestation
Fear	Withdrawal
Rejection	Subversion

THE RESEARCH STRATEGY

The contemporary cities chosen for investigation had to have growing black majorities, popularly elected black mayors, and transitions of sufficient recency to guarantee that black rule had not become entirely routine. These criteria eliminated places like Los Angeles and Raleigh for their lack of a black majority, several smaller cities for the fact that their mayors are elected by the city council, and Gary and Newark for the relative longevity of their black regimes. Detroit and Atlanta, whose black mayors came to power in the fall of 1973, not only met all these criteria when the study commenced in 1975, but also offered the following advantages:

1. Both are large cities with a variety of life-styles, classes, ethnic groups, and political groups.
2. Both have good local newspapers as well as permanent *New York Times* correspondents.
3. Each city is ringed by a substantial metropolitan area, providing viable possibilities for elite "exit" as well as for metropolitan reform.
4. The two cities offer a variety of important contrasts in terms of population composition, history, economic base, and regional location.

The information on which the analysis of white elite adjustment to displacement is based was gathered from a wide range of written sources, including newspapers, magazines, and various public and private reports and documents, as well as from a series of interviews conducted with white elites in each city. Nearly 80 initial interviews were carried out, half in Atlanta during the summer of 1975 and half in Detroit during the first 4 months of 1976. During follow-up visits to both cities in the spring of 1978, more than a dozen additional interviews were completed.[16]

Although both elites and masses undoubtedly need to adjust to the loss of power by their group, the problem is not only more severe for those who had actually participated in governing the city or influencing its life and development, but the implications of their adjustment are more important for the new regime and the city by virtue of their continued control of certain resources. Thus the study focuses on white elites in the two cities who occupy positions from which they control major political or economic resources ranging from the control of the jobs and tax base inherent in high corporate executive positions, to the wherewithal to finance campaigns, to prestige, political followings, expertise, organizational position, and information. People with such resources were to be found primarily in the political, business, and media sectors. They were identified in part by a survey of contemporary newspaper and periodical sources, through discussions with informants (mainly academics and journalists in each city), and by nomination by other

---

[16]References for interview material are cited in the text according to a code (e.g., I-137) rather than by name in order to preserve the anonymity of my respondents. See Appendix A for a list of people interviewed.

interviewees. Others were selected by virtue of the formal roles they played. Thus, a decision was made from the outset to include the names of Chamber of Commerce officials, newspaper editors, major white officeholders (past and present), and civic association leaders, whether their names subsequently appeared in the written sources or were mentioned during the interviewing.

The major bias in such a procedure, perhaps, is that those finally selected tended to have either a visible record of local civic or political involvement, or a reputation for being influential in local affairs. Thus some individuals, particularly certain corporation executives oriented toward the national arena, were less likely to be interviewed, even though they were capable of making decisions that could influence the economic health of the city. In neither city did the pool of such people appear to be large, however, for in both places the norm of local civic involvement by the business community was unusually strong.

The final sample in each city was in no sense a random representation of some indefinable universe of elites. Instead, the sampling strategy was to select elites who controlled a variety of different types of resources of potential importance to a mayor, and who moved in various (often autonomous) spheres of activity. Although those interviewed were people whose names tended to be mentioned repeatedly by the various sources through which the sample was compiled, no claim is made that they constituted a "power elite" either before or after the transition to black rule. The identification of the "most powerful" people in each city or the "top rulers" was not the intention of the study, nor was evidence of influence in particular decisions a criterion for inclusion on the list of those interviewed.

Such a sampling procedure—or indeed any sampling procedure—cannot guarantee that the investigator will identify the people who control most of the important resources in a community; but employing the technique with care will result in the inclusion of at least most of those in positions of formal civic and political authority, many behind-the-scenes people (who appear in the sample by nomination), and those business, civic, and political activists who have a record of using their resources in local affairs. The sampling procedure is less likely to catch those individuals who may control major resources but have never used them. Despite this limitation, the final samples in each city, of whom approximately 80% were interviewed, certainly included a large number of people who controlled truly significant resources, the disposition of which would make a difference to how any mayor managed to govern.

In analyzing and reporting on the interview data, I have, in most instances, avoided offering percentages or sums. The reasons are several.

For one thing the interviews, although structured, were *elite* interviews, that is, "something which sounds like a discussion but is really a quasi-monologue stimulated by understanding comments [Dexter, 1970, p. 56]." Because those interviewed did not occupy comparable positions or control easily comparable resources, they were not all asked the same questions. Such a procedure makes sense when it is understood that those interviewed were important not so much as more or less interchangeable representatives of a universe of elites but as individuals. Each person had unique information to offer and often had control over unique or at least very important resources. Thus the opinions of these individuals are important in the singular. Summing them for the appearance of precision was normally unnecessary. Furthermore, their opinions may not be of comparable significance. One cannot assume, for example, that a backbench alderman's optimism about the future of the city offsets the pessimism of a newspaper editor; yet both are worth interviewing, for each controls or influences the disposition of important resources. Nevertheless, it has occasionally been necessary to make general statements about the structure of opinion, to describe majority and minority views, to suggest modalities. For this I have simply exercised a judgment informed by my experience in the two cities. If I have erred here, it is that I have been unusually conservative in my conclusions.

The analysis of the process of ethnoracial political transition begins in Chapter 2 with a case study of Yankee adjustment to the loss of political hegemony in Boston in the half century after 1884. The purpose of this historical chapter is to seek potential lessons that might inform the investigation of the contemporary cases of transition. Chapter 3 offers a glimpse of the research settings of Detroit and Atlanta. In the next two chapters the psychological adjustments (Chapter 4) and the strategic adjustments (Chapter 5) of white elites in the two modern cities are described and analyzed. Chapter 6 offers a brief analysis of metropolitan reform efforts in the two cities as a way of illustrating the limits of a subversive strategy. Chapters 7, 8, and 9 probe explanations for the nature of the white adjustment, which is strikingly acquiescent, benign, cooperative, and accepting in both modern cities.

2

Transition to Irish Rule in Boston, 1884–1933: A Case Study*

Boston Yankees in the last decades of the nineteenth century anticipated the loss of political power to the Irish with apprehension and a good deal of bitterness.[1] In Marquand's novel (1937) of the decline of the Brahmins, Thomas Apley sadly lectures his son: "It is not a pleasant thing for me to feel that the Irish are going to run the affairs of this city, and I do not see anyone in your generation who has the force and skill to guide them [p. 149]."

The outlines of the process of ethnoracial transition in Boston are simple enough. The foundations of Irish electoral power were laid in the great immigration of the 1840s from the Emerald Isle, and in the ranks of cheap Irish labor attracted to Boston by the industrializaton of New England in the 1850s. The Irish emerged as a serious political force in the

---

*This chapter is reprinted in revised form with permission from the *Political Science Quarterly* 93 (Summer 1978): 217–239.

[1]*Yankee* is a term used to refer to an ethnic group whose origins are predominantly Anglo-Saxon, whose religion is predominantly Protestant, and whose shared heritage lies generally in the colonial past. Such a definition draws on the notion of "ethnic group" proposed by Schermerhorn (1970, p. 12). In this chapter I use *Yankee* interchangeably with the term *Brahmin*, after Oliver Wendell Holmes's designation of the New England Yankee elite (see Solomon, 1956, p. 3).

decades after the Civil War. In 1883 Senator George Frisbie Hoar wrote grimly to Henry Cabot Lodge, "Unless we can break this compact foreign vote, we are gone, and the grand chapter of the old Massachusetts history is closed [Barbrook, 1973, p. 16]." By 1884 the Irish of Boston were able to elect one of their own to the mayoralty. Symbols of the transition in process followed closely: In 1885 Patrick Maguire, the most prominent Boston ward boss, became the first Irish Catholic to deliver the city's Fourth of July oration (Blodgett, 1966, p. 145), and in 1892 the city closed the public library on St. Patrick's Day, enraging the Protestant community (Solomon, 1956, p. 88). Through the first years of the new century each fresh Irish political triumph elicited Yankee eulogies to the "old Boston." By 1914, when James Michael Curley defeated Thomas J. Kenny in the first mayoral contest involving two Irish Democrats, Yankee political prospects were at a virtual end. In 1930 a memorial volume marking the tercentenary of Boston's founding was able to observe with thorough dispassion that control of municipal government had passed out of the hands of the "Colonial Americans" and into those of the "newer races" (Herlihy, 1932, p. 83).

Essentially, the transition has been complete (see Table 2.1). Of the 16 men who sat in city hall between 1884 and 1979, 9 have been of Irish descent. Since 1901 the Irish have controlled the mayor's office for all but 10 years, losing it only to George Hibbard for a 2-year term in 1907 in the aftermath of fiscal scandal; to Andrew Peters in 1917; and to Malcolm Nichols in 1925 when three Irish candidates split the Democratic vote. However, it took nearly half a century for the process to be played out. Hugh O'Brien's election in December of 1884 marked a turning point in the conflict between Yankee and Irishman, but neither that election nor James Curley's 1914 win served as a *coup de grace* to Yankee power and aspirations. The Yankee descent to minority status was marked by a series of holding actions and temporary periods of political revival.

The patterns of Yankee adjustment were both varied and complex. The Yankees were acutely aware of the process of ethnoracial transition: Civic leaders were not only self-consciously caught up in it and fearful of its consequences, but they were also intensely reflective about their own roles as members of a threatened caste. Psychological responses ranged from an initial sense of fear to eventual acceptance. For most Yankees a stern sense of Brahmin dignity, a characteristic ability to apprehend and confront reality, and a commitment to the integrity of the Brahmin tradition foreclosed those psychological responses labelled rejection, rationalization, and satisfaction. The basic pattern of strategic response shows a shift from cooperation to withdrawal, underlaid by persistent attempts at contestation through 1925 and periodic efforts at subversion.

TABLE 2.1
Boston Mayors and Runners-up, 1884–1975

	Winner		Runner-up	
1884	Hugh O'Brien	(Irish, Democrat)	Augustus P. Martin	(Yankee, Republican)
1885	Hugh O'Brien	(I,D)	John M. Clark	(Y,R)
1886	Hugh O'Brien	(I,D)	Thomas Hart	(Y,R)
1887	Hugh O'Brien	(I,D)	Thomas Hart	(Y,R)
1888	Thomas Hart	(Y,R)	Hugh O'Brien	(I,D)
1889	Thomas Hart	(Y,R)	Owen Galvin	(I,R)
1890	Nathan Matthews	(Y,D)	Moody Merrill	(Y,R)
1891	Nathan Matthews	(Y,D)	Horace Allen	(Y,R)
1892	Nathan Matthews	(Y,D)	Homer Rogers	(Y,R)
1893	Nathan Matthews	(Y,D)	Thomas Hart	(Y,R)
1894	Edwin Curtis	(Y,R)	Francis Peabody	(Y,D)
1895[a]	Josiah Quincy	(Y,D)	Edwin Curtis	(Y,R)
1897	Josiah Quincy	(Y,D)	Edwin Curtis	(Y,R)
1899	Thomas Hart	(Y,R)	Patrick Collins	(I,D)
1901	Patrick Collins	(I,D)	Thomas Hart	(Y,R)
1903	Patrick Collins	(I,D)	George Swallow	(Y,R)
1905	John F. Fitzgerald	(I,D)	Louis Frothingham	(Y,R)
1907	George Hibbard	(Y,R)	John F. Fitzgerald	(I,D)
1910[b]	John F. Fitzgerald	(I,D)	James Storrow	(Y,D)
1914	James M. Curley	(I,D)	Thomas Kenny[c]	(I,D)
1917	Andrew Peters	(Y,D)	James M. Curley	(I,D)
1921	James M. Curley	(I,D)	John R. Murphy[c]	(I,D)
1925	Malcolm Nichols	(Y,R)	Theodore Glynn	(I,D)
1929	James M. Curley	(I,D)	Frederick Mansfield[c]	(I,D)
1933	Frederick Mansfield[c]	(I,D)	Malcolm Nichols	(Y,R)
1937	Maurice Tobin	(I,D)	James M. Curley[d]	(I,D)
1941[e]	Maurice Tobin	(I,D)	James M. Curley	(I,D)
1945	James M. Curley	(I,D)	John Kerrigan	(I,D)
1949	John B. Hynes	(I,D)	James M. Curley	(I,D)
1951[f]	John B. Hynes	(I,D)	James M. Curley	(I,D)
1955	John B. Hynes	(I,D)	John Powers	(I,D)
1959	John Collins	(I,D)	John Powers	(I,D)
1963	John Collins	(I,D)	Gabriel Piemonte	(Italian, D)
1967	Kevin White	(I,D)	Louise Day Hicks	(I,D)
1971	Kevin White	(I,D)	Louise Day Hicks	(I,D)
1975	Kevin White	(I,D)	Joseph Timilty	(I,D)

Source: From Eisinger (1978). Reprinted with permission from the *Political Science Quarterly, 93* (Summer 1978), 221.

[a] Two-year terms came in.

[b] First nonpartisan electon; 4-year terms instituted.

[c] Irish Democrats endorsed by the Good Government Association (GGA), dissolved December 4, 1933.

[d] Malcolm Nichols ran third in this election.

[e] First time since 1919 charter that a mayor could succeed himself.

[f] Runoff system instituted.

Prior to 1900 Yankee "Mugwumps" established a deliberate policy of cooperation with the Irish Democracy in Boston politics. At the same time, Yankee Republicans were slowly withdrawing from the local political arena, turning their major energies to state and national politics, club life, and private charities. But after the turn of the century withdrawal took a new form as the Mugwumps (whose coalition with the Irish dissolved during the depression of the mid-1890s) and the Republicans gradually joined forces in the effort to restrict Southern and Eastern European immigration, diverting energies once devoted to confronting the Irish challenge. Urban politics were not entirely abandoned, however. Yankee Republicans contested 6 of the 11 elections between 1901 and 1933, and in league with Yankee Democrats they made their weight felt in other elections through various reform organizations, principally the Good Government Association; but after 1933, no candidate of Yankee stock, Democrat or Republican, appeared among the top contenders for city hall. During much of this period it was hoped that institutional structural changes, such as charter reform and metropolitanization, would forestall the growth of Irish strength and even make possible a reassertion of Yankee dominance.

Yankee adjustments to the loss of hegemony in the local political sphere were sustained by broad sociopolitical movements of the times. The conflict in Boston between the two ethnic groups was fought on terms understood not only elsewhere in New England but also in the middle and western United States. Although the urgency, flavor, and significance of the struggle were cast in peculiarly New England terms, it was nevertheless, in some measure, a microcosm of the broader contest over reform and immigration. Boston was not simply the last threatened bastion of New England culture, nor an isolated and unique locus of urban conflict, but rather a place where variations of broader national forces were in conflict. Thus what we learn from Boston can be applied to American society as a whole in the decades around the turn of the century.

A detailed examination of transition in Boston may also throw light on contemporary trends, for there appear to be important similarities between the situations of the Irish and the blacks and their respective achievements of political power at the local level. This does not mean that black Americans in the 1970s may anticipate the level of acceptance and political integration achieved by the Irish in the 1930s, or that the historical experiences of blacks are comparable to those of the Irish. The ineradicable residues of racial prejudice and the racial caste system serve to set the black struggle in a category by itself. Yet there are parallels in the rise of both groups to political power.

Both the Boston Irish in the years prior to the Civil War and urban black Americans in the period after World War I occupied positions as least-favored groups in ethnically stratified cities. They were both overwhelmingly concentrated in the unskilled and marginal ranks of the labor force and subject to severe discrimination in terms of employment opportunities and earnings (Handlin, 1941/1974, pp. 57, 60; Levine, 1966, p. 59). In Boston in the years around the Civil War the Irish, who were popularly blamed for most crimes of violence, dominated the prison population just as blacks in large urban areas have since their move *en masse* to northern cities (Handlin, p. 257). Irish and black residential patterns during these respective periods displayed more extreme levels of segregation than were imposed on any other ethnic group (Handlin, p. 91; Lieberson, 1963; Taeuber and Taeuber, 1965), and both groups have been subject to the same stereotypical characterization as "lazy, shiftless, dirty, savage" (Greeley, 1972, p. 225).

Nevertheless, the Irish in the post-Civil War decades and the blacks in the 1960s began to achieve urban political power through the exercise of bloc voting in particular cities, although their respective routes to political mobilization differed. When the political breakthough to control of city hall occurred in Boston, Atlanta, and Detroit, neither group had achieved the formation of a substantial middle class; thus the disposition of important middle-class resources controlled by members of the displaced groups became critically important to the new victors. How the new black urban politicians have sought to gain access to these resources is a subject for investigation. It is a matter of record, however, that Irish machines harnessed middle class money through alliances with business interests, offering a variety of favors in exchange for financial sustenance.[2]

Focusing on the Boston case study allows us to explore the process of ethnoracial transition as it was played out over time. What becomes important in historical perspective is not simply the modes of adjustment of the displaced group at any given point in time, but the changes in modes of adjustment. Because of the apparent basic similarities in the forces that led to transition in Boston and the contemporary cities, as well as the generally analogous character of the groups in conflict, the Boston case can be used to establish benchmarks to watch for in the investigation of the transition process in Detroit and Atlanta. In addition, the historical case study permits the generation of predictive hypotheses regarding the course of transition in modern cities undergoing changes in racial balance, while in effect controlling for the unique racial factors at work.

[2]For one of the latest arguments to this effect, see Shefter (1976).

THE RISE OF THE IRISH IN BOSTON POLITICS: A BRIEF SURVEY

Yankee Boston had been quick to perceive that the early Irish immigration was a challenge to its secure society. Yet as streams of impoverished Irish peasants made their way from the packet landings in the Maritime Provinces down the New England coast to the city, the problem that troubled the Brahmin elite in those antebellum years was less the threat of displacement than, as Charles Eliot Norton put it, the difficulty of absorbing "a people so long misgoverned . . . a race foreign to our own [Solomon, 1956, p. 12]."

Indeed the Irish were a very different people and, what was worse in Brahmin eyes, an uneducated and reactionary people. They shared none of the Yankee reformist passions. Irish Catholicism in the American idiom provided no sustained impulse for the idea of liberal education, nor did it supply ideological grounds for supporting abolition, women's rights, or public education. In addition, Irish mores precluded sympathy for the temperance movement (Green, 1966, p. 46). As Boston's Irish population continued to increase (by 1840, 46% of the city's population was foreign born, nearly all of it Irish; Ward, 1963, p. 339), the carefully wrought culture of New England's hub seemed genuinely threatened.

With only a few exceptions, however, Boston's Brahmins refused to endorse the Know-Nothing movement that developed in the decade between 1845 and 1855. The brief domination of Massachusetts politics by this virulently anti-Catholic party shocked Brahmin democratic sensibilities, and led them to reaffirm their faith in the essential correctness of an open immigration policy.

Meanwhile, Irish Catholics began to make political inroads throughout Massachusetts. Boston elected its first Irish Catholic common councilman in 1857; in 1870 it had its first Irish alderman. City voters sent the state's first Irishman to Congress in 1882, and then in 1884 the Irish captured the mayoralty. As early as 1895—and probably several years before that—people of Irish descent constituted a substantial majority in Boston, a position they maintained at least into the 1960s.[3] Brahmin ideals seemed suddenly self-defeating.

As the campaign for mayor began in 1884 it was clear to many that

---

[3]Determining the ethnic composition of the city's population is a difficult task. Census data report only the first and second generation ethnics. Blodgett (1966, p. 149) estimates the total Irish population at more than 60% in 1895. Others offer estimates that suggest that the Irish maintained their majority status at least through the 1960s (see Shannon, 1963, p. 186; Barbrook, 1973, p. 22; Banfield, 1965, p. 38).

the end of uninterrupted Yankee rule was near. Mayor Augustus P. Martin, a Republican Mugwump, had beaten Hugh O'Brien, a five-time alderman, in the 1883 race by only 1500 votes. The Brahmins anticipated the end with foreboding. The principal mouthpiece of Brahmin sentiment, *The Boston Evening Transcript*, quoted a prominent Boston Republican's remark that "the day has gone by when a Republican can be elected on party principles in Boston [20 Nov. 1884]," and intoned, "The cause of good government in Boston, not only for this year but for an indefinite time to come, is bound up with the success of General Martin and the non-partisan movement [5 Dec. 1884]."[4]

It is important to understand at the outset that although Boston's Yankee elite was in no sense monolithic in the way its members adjusted to Irish power, the broad concern they felt over displacement and the subsequent measures some of them took to regain power must be regarded as indications of the general importance of political control in the nexus of Brahmin values. The Yankee response also suggests that they had no immediate alternative means of exercising collective power in the society as satisfactory as the political system. The Brahmins were men who nourished a vision of the public good to be realized not principally through the good works of the economically well-off individual in the marketplace but through the vehicle of the polity and the duty of public service.[5]

The O'Brien–Martin race of 1884 solidified the terms in which much ethnic conflict was to be understood in Boston. In Brahmin public discourse the issue of ethnic struggle came generally to be sublimated to the broader contest between reform and machine. Local Republicanism, pictured as the domain of the "good people" of Boston, was equated with nonpartisanship, whereas the Democratic "ring" came to stand for the vehicle of Irish power *(BET*, 25 Nov. 1884).[6] What is significant is not so much the sense of nonpartisan virtue with which Yankee Republicans invested themselves, but the fact that public exploitation of the ethnic dimensions of the struggle in progress became almost from the start virtually unacceptable to the Brahmins. "It has been supposed," wrote the *Transcript* in a double-edged rebuke during O'Brien's campaign for a second term in city hall, "that the party dominant in this city was Amer-

---

[4]On Yankee apprehensions over the impending loss of political power, see also Shannon (1963, pp. 208–209) and McFarland (1975, p. 135).

[5]See, for example, the essays by the Mugwump leader Moorfield Storey (1889).

[6]Such an equation was common throughout the period of transition. For example, commenting on the campaign of 1905, the *Transcript* wrote, "If Mr. Frothingham is elected he will devote himself to the city; if Mr. Fitzgerald, he will devote the city to himself [25 Nov. 1905]."

ican, not Irish-American. . . . The effort made in some quarters to discredit [O'Brien] because of his race and religious belief is despicable. It shows a very superficial knowledge of true republicanism for any race in our community to boast that it governs Boston as such [4 Dec. 1885]." Brahmin animosity toward the Irish did not, of course, disappear, but it was very quickly largely masked.

O'Brien beat General Martin by more than 3400 votes out of nearly 52,000 cast to win his first term. *The Boston Post* (10 Dec. 1884) called the result regrettable but necessary. *The Boston Herald* of the same date, noting O'Brien's Irish Catholic background, proudly claimed that the election result demonstrated Boston's lack of prejudice.[7] O'Brien promptly disclaimed any association with the "ring" *(BET,* 10 Dec. 1884) and embarked upon a year in office that had all the trappings of a reform adminsitration. The *Nation* (1885, p. 124) soon compared him to Grover Cleveland. Even the *Transcript,* which had called O'Brien the candidate of the bosses, eventually praised him for his attentiveness to voters who wanted only "an efficient, economical and upright city government [2 Dec. 1885]."

O'Brien went on to win three more single-year terms. Although Yankee tolerance of O'Brien began to run short by the end of his second term, his achievement in the end was to set a standard of conciliation that dominated Boston politics until the end of the century, easing the city toward Irish control in a most gentle way (Blodgett, 1966, p. 61).

There were, to be sure, early periods of difficulty for the Irish on the road to transition. A combination of factional fighting within Democratic ranks—a chronic problem—and the crystallization of anti-Catholic feeling in Boston led to O'Brien's defeat in 1888. It is notable, however, that Boston's established Yankee aristocracy neither raised nor sought to exploit with vigor the issue of Irish Catholicism. Anti-Catholic sentiment was most forcefully articulated by an organization—formed in 1887 by working class Canadian immigrants—that called itself the British-American Society. Expressing concern over the survival of the public school system in an increasingly Catholic city, the British-American Society became a minor factor for a brief time in Boston's mayoral contests. The society enjoyed the tacit support of another newly formed group, the loosely organized Citizen's Committee, which drew its membership from disaffected Yankee Democrats and some Republicans. The Citizen's Committee, which O'Brien believed was the source of a widely distributed anti-Catholic tract during the 1887 campaign *(BET,* 14 Oct.

---

[7]Detroit and Atlanta newspapers indulged in similar self-congratulation when black mayors were elected.

1887), sought unsuccessfully to run a third-party candidate in 1887. In 1888 and 1889 it joined in the successful Republican efforts to win with Thomas Hart as the standard bearer.[8]

Hart's consecutive victories over O'Brien and a young Irishman named Owen Galvin led the city's Irish Democratic leaders to reassess their strategy. Although Yankee Democrats had essentially conceded control of Boston to the Irish in order to hold the urban wing of the party in state contests, Patrick Maguire persuaded his fellow Irish party leaders to relinquish the plum of the mayoralty in order to dampen rising anti-Catholicism (Blodgett, 1966, pp. 158–170). Accordingly, he set out to find a Yankee Democrat to run, eventually selecting Nathan Matthews, the organizer of Boston's Young Men's Democratic clubs and a personal friend.

The *Evening Transcript* (22 Nov. and 1 Dec. 1890) was not impressed by the choice, charging that Matthews was a partisan Democrat and a candidate of the bosses. But Maguire's strategy was aided by the Republican nomination of a colorless former state legislator named Moody Merrill, whose defeat of Thomas Hart in the GOP nominating convention divided the party. Independent Democrats, primarily Yankees who had supported Hart, were drawn back to the fold and helped to produce a landslide victory for Matthews in 1890, vindicating Maguire's calculations (4 and 10 Dec.). In later mayoral races Matthews added to his coalition by gaining the support of the normally Republican businessmen of the Citizens' Association (11 Dec. 1893).

Matthews was one of a number of young Boston Yankees who came of political age with the Mugwump movement after 1884. Unlike the Republican defectors who spurred the Mugwump break, he entered politics as a Democrat. Realizing the party's dependence on Irish voting strength in statewide contests, Matthews and other young Democratic Mugwumps embarked on a period of cooperation with the Boston Irish (Blodgett, 1966, p. 85).

The coalition Matthews and Maguire forged to hold the mayoralty in Democratic hands began to come apart, however, in the economic depression of 1893, in particular over differences of opinion as to the

---

[8]In his autobiography, James M. Curley (1957, p. 44) calls Hart a "renegade Irishman," referring to his Republicanism. There is no evidence I could find, however, that suggests that Hart was indeed an Irishman or that he was perceived as such. Hart took tacit advantage of anti-Catholic sentiment in the mayoral campaign of 1887. His speeches were filled in each of his many campaigns with appeals to Yankee business and reform instincts, and he shared Yankee fears of a Catholic "threat" to the public school system. He sounded always like the banker he was. In any event, Patrick Collins, elected first in 1901, is widely regarded as the second Irish mayor of Boston (see, for example, Peters, 1930, p. 77). If Hart was indeed of Irish descent, he concealed it very well. In this study I shall consider him a Yankee.

municipal government's obligation to the large pool of predominantly Irish jobless men in the city. Maguire finally broke with Matthews in 1894. In seeking a replacement in keeping with his strategy of reserving the head of the local ticket for a Protestant, Maguire hit upon Francis Peabody, a Yankee lawyer of impeccable connections but a man with no experience in public or party service (BET, 20 Nov. 1894). It proved a ludicrous choice. Unbelieving Irish ward bosses sat out the election of 1894, enabling Edwin Curtis, a Republican, to win a year in city hall.

Although Maguire faced growing discontent among an impatient corps of ward bosses, which included the young Martin Lomasney and John F. Fitzgerald, Irish Democracy nevertheless sought one more time to follow a Protestant strategy in the mayoral race of 1895, supporting Josiah Quincy, two of whose forebears of the same name had served as mayors of Boston. Quincy had earned his credentials as a party official and campaign manager in state and national politics, but what interested the Irish in particular was his reputation in the state for a generous patronage policy (BET, 14 Nov. 1895). Once elected to the first of what were now 2-year terms, Quincy established a backroom "Board of Strategy," composed mainly of Irish party politicians in the city, to allocate patronage and to plan electoral strategy.

In 1896 Pat Maguire died. By force of will he had imposed a thin veneer of unity on the Boston Democratic party. With his departure the fragmented structure of the party became visibly pronounced. Warring factions were unable to agree on a candidate to succeed Quincy in 1899, throwing the election to Thomas Hart (BET, 21 Nov. and 6 Dec. 1899).

Hart's election marked the last time that a candidate backed by the Yankees won more than 50% of the vote (see Figure 2.1). From the turn of the century on, Yankee fortunes depended very heavily on the extent to which the Irish Democrats failed to hold their warring factions together. Yankee political power in these years was aggregated and exercised mainly through the Good Government Association, a reform group put together, ironically, by Louis Brandeis, a Mugwump, in 1903.[9] From that year until 1933, when the GGA was dissolved, three men endorsed by the reform organization won the mayoralty: Andrew Peters, a Yankee Democrat, in 1917; Malcolm Nichols, a Yankee Republican, in 1925; and Frederick Mansfield, an Irish Democrat and enemy of James Curley, in 1933. Each of these three men won substantially less than half the votes, as numerous Irish contenders, divided largely by the nearly 20-year

[9]The GGA was once described by Curley with some accuracy as "that select and exclusive body of social bounders in the Back Bay [NYT, 5 Nov. 1929]."

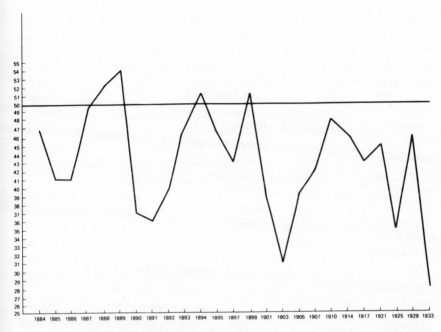

Figure. 2.1. Fluctuation of the vote for Yankee-backed mayoral candidates, Boston, 1884–1933. [From Eisinger (1978). Reprinted with permission from the *Political Science Quarterly* 93 (Summer 1978), 226.]

feud between Curley and John F. Fitzgerald, fragmented the Irish vote. In 1929, "Honey Fitz" and Curley united the party, and from that year on the mayoralty has belonged exclusively to one or another Irishman.

PSYCHOLOGICAL ADJUSTMENT TO ETHNORACIAL TRANSITION

The psychology of displacement in Boston may be reconstructed from the writings of people who, as best we can tell, reflected the concerns of their class and age. Perhaps more than anywhere else in America, cultural standards and social behavior in nineteenth-century Boston were shaped and dominated by a distinct literary tradition. This was in part a product of the high degree of overlap between doers and writers in Boston civic affairs. Academic men in particular were singularly prominent among the city's influential forces. These journal keepers, letter writers, and pamphleteers set the standards, especially after the pre-

Civil War heyday of the Brahmin merchant class, by which the "proper Bostonian" thought, lived, and died. Thus, what they wrote can, I believe, generally be taken as "emblematic" of the larger culture, that is, as relatively accurate mirrors of the elite psychology of the times as well as powerful attempts to shape it. [10]

America, as Yankee New Englanders understood it, was clearly under assault in the last decades of the nineteenth century. Henry Adams sought a spiritual retreat in the Mariolatry of gothic France, but a number of more practical members of the Brahmin class simply moved to rural Brookline. Immigration trends, combined with the ugliness of industrialization, produced a mood of despair and depression (Handlin, 1941/1974, p. 221). This was translated easily into fear of the Irish challenge, which "in race conscious America," Abrams (1964) observes, "meant only barely less than a challenge to civilization itself [p. 133]."

I have defined *fear* in this study as a mode of psychological response that stresses a concern for the survival of a particular life-style and its associated value structure in the face of threat. In a number of Brahmin writers, fear of the Irish produced an almost wistful sense of loss. The Harvard art historian, Charles Eliot Norton, wrote in a letter in 1897, "I fancy that there has never been a community on a higher and pleasanter level than that of New England during the first thirty years of this century, before the coming in of Jacksonian Democracy, and the invasion of the Irish [Norton and Howe, 1913, p. 254]." And Gamaliel Bradford, a descendant of the governor of Plymouth Colony and a prominent Mugwump, wrote in his journal in the same year, "This afternoon, town meeting. Always entertains me. . . . In its origin, for hard-headed, sober, deliberate New England farmers it was an excellent institution; but now that the pyrotechnic Celt has got hold of it, it is in a rapid state of decline, I imagine [Brooks, 1933, p. 85]."

Bradford, virtually alone among the Brahmins, mixed his fear with a sense of personal revulsion, confessing to a feeling of "terror" when he was brought into contact with the "low-browed, dirty, hard-handed animal sons of toil" from Ireland (Brooks, 1933, p. 85). Others, however, regarded the Irish less in terms of their threat to personal sensibilities or to New England culture than as a challenge to what they saw as more generalized American values. Moorfield Storey, a Mugwump leader and prominent Boston attorney, worried about the survival of "free government." In an essay entitled *Politics as a Duty and as a Career*, published in 1889, he wrote:

[10] Green (1966, pp. 10, 12, 26, 35) uses the term *emblems* of the culture in his work and makes a similar argument.

A republic cannot succeed if it becomes an oligarchy of "bosses." . . .
The immigration of every year adds to the mass of poverty and igno-
rance in our country. [The immigrants are unfit] to take part in our
political contests, yet in a few years they become citizens and their
votes in the ballot box count as much as our own [pp. 4-5].

Henry Cabot Lodge, a prime mover in the U.S. Senate to restrict the
flow and origins of immigration, blamed the immigrants (by whom he
chiefly meant the Irish) for the rise of professional politicans, municipal
corruption, city debts, and inefficient urban administration, particularly
in Boston (1902, pp. 198-99).

Brahmin fears of the Irish in these years, however, although apparent, were remarkably tempered and ultimately limited by a larger and
continuing commitment to some democratic ideal. Norton affirmed his
faith in democracy in a letter to James Russell Lowell in 1884, although
he worried that "it may work ignobly, ignorantly, and brutally [Blodgett,
1966, p. 33]." Years later Gamaliel Bradford reports on the continuation
of his "old controversy" with George Frisbie Hoar, U.S. Senator from
Massachusetts:

He is sceptical about the future of democracy. So am I. Then . . . he
exclaims, "After all, it would be better if we had a monarchy." And
against this I protest . . . with all the energy of my being. Democracy is
a great experiment, an enormously difficult experiment [Brooks, 1933,
p. 112].

Hoar himself, who may on occasion have doubted the wisdom of
democracy, nevertheless vociferously attacked the anti-Irish prejudice
of the American Protective Association in the 1890s. In a speech of 1895
he declared, "I have no patience or tolerance with the spirit which
would excite religious strife. . . . This nation is a composite [1903, p.
283]." In light of such beliefs it is no surprise that the sense of loss
experienced by New England's Yankee elite never found expression in
reactionary antidemocracy.[11] Bitterness and doubt were instead mostly
contained (Solomon, 1956, p. 153), giving rise to interior personal strug-
gles between values and impulses that laid the basis for eventual acceptance of the new political reality.

In part, Yankee Boston managed the transition from fear to uneasy
acceptance of Irish power by initially patronizing the new majority in a

[11]To be sure, a number of Brahmins participated as leading figures in the movement to
restrict immigration. Yet these efforts, which were justified in part as necessary to save
American democracy, never sought to call into question the principles of popular government, majoritarianism, or civil liberties. Nor did they seek to subvert existing bases of
immigrant power through antidemocratic means.

way that seemed deliberately designed to obscure the depths of change in the political order. After a period of social controversy, the Irish domestic had become an acceptable fixture in Brahmin households (Solomon, 1956, p. 153), nourishing caste thinking but also softening its effects. In a volume of boyhood recollections, Samuel Eliot Morison (1962) remembers that "the family was . . . down on corrupt Irish politicians," but that "in the close daily contact as we were with Irish maids and workmen, we could believe no ill of the Irish [p. 64]." In a related vein Bliss Perry (1921) could describe the archetypal Brahmin, Henry Lee Higginson, stockbroker and founder of the Boston Symphony Orchestra, as an "old army officer" who knew how "to 'hit it off' with Irishman and Hebrew, Negro and Italian [p. 454]."

The path to acceptance of Irish majority power was opened early in the contest between the two ethnic groups by a small number of Brahmin reformers committed to a notion of cultural pluralism. Along with men like George F. Hoar, C. W. Eliot, and the Reverend Edward Everett Hale, Thomas Wentworth Higginson emerged as an early and articulate Yankee defender of the Irish. Higginson, who had commanded a black regiment during the Civil War, had already distinguished himself as an abolitionist. In 1880 he stood for election to the state legislature as a Republican and won. It was here that his position on the Irish began to develop. In the mid-1880s he wrote that "a due proportion of Irish admixture will help, not hinder American society." In 1888 he led the effort in the legislature to defeat the School Inspection Bill, a thinly veiled assault on the newly established parochial school system in the state. In 1891 at a memorial service for the Irish American writer John Boyle O'Reilly, Higginson argued that America had failed "if it was only large enough to furnish a safe and convenient place for the descendants of Puritans and Anglo-Saxons, leaving Irishmen and Catholics outside [Edelstein, 1968, p. 370–372]."

Such sympathetic views were not, of course, broadly characteristic of Brahmin sentiment, but at the turn of the century there occurred a more general reevaluation of the Irish stimulated by the contrast posed by the new immigration from Southern and Eastern Europe. Boston social workers, struck by the apparent intractability of the more exotic non-English-speaking newcomers to the city, suddenly began to speak of the "race trait adaptability" of the Irish (Solomon, 1956, p. 154). Even prominent immigration restrictionists such as Francis Amasa Walker and Henry Cabot Lodge began to soften their views of the Irish. On the basis of his recurrent census analyses beginning in 1870, Walker, an economist and president of the Massachusetts Institute of Technology, had argued that Irish, German, and Scandinavian fecundity was slowly overwhelming the less fertile but superior Anglo-Saxon "race." Immigration restriction, in this context, became a matter of racial preservation. In the early 1890s, however, the shifting composition of immigrant newcomers in favor of Italians, Poles, and Russians cast the Irish in a new light. Walker decided that the Irish had worked hard, raising their standard of living. More important, the republic seemed to have survived the Irish, but it was not at all clear that it could do so in the face of the newer arrivals (Solomon, 1956, pp. 70–79).

Admiration for Irish socioeconomic mobility was reflected in the writings of others in this period. John D. Long, a former governor of Massachusetts (1880–1882) and a prominent national Republican, wrote in his journal in 1906:

> In the evening, I went to a local Hibernian meeting [in Hingham], held in our Loring Hall, in honor of St. Patrick. . . . My talk was on the Irishmen of Hingham. The generation of Irishmen who came here, 60 or 70 years ago, were a good stock—just as good as our Pilgrim ancestry, and very much like them, in some ways. They brought nothing but their hands. The children of these first-comers, the present generation, . . . turn to professional or business life. They run to politics. The severe manual, rough toil, which their fathers did, is now done by Italians, Poles, or other later arrivals [Long, 1956, p. 291].

The Brahmin reevaluation of the Irish is most strikingly illustrated in the transformation of Henry Cabot Lodge. In 1881 Lodge had written in *A Short History of the English Colonies in America* that the Irish were "a very undesirable addition at that period. . . . They were a hard-drinking, idle, quarrelsome and disorderly class . . . and did much to give to government and to politics the character for weakness and turbulence [Saveth, 1948, pp. 56–57]." Yet as the Irish became an increasingly large component of Lodge's Senate constituency, he began to temper his views of them (a race "closely associated with the English-speaking people") as he pressed for immigration restriction. In contrast to the Southern and Eastern Europeans, he argued before the Boston City Club in 1909, the Irish "presented no difficulties of assimilation, and they adopted and sustained our system of government as easily as the people of earlier settlement [p. 58]."

The change in Lodge's views was not merely a product of political calculation. In 1890 he had begun research on an article entitled "The Distribution of Ability in the United States" in which he sought to classify the "race" backgrounds of the 15,000 people listed in Appleton's *Encyclopedia of American Biography*. His analysis showed that the Irish were disproportionately represented among the notable Americans included in the *Encyclopedia*. A Lodge biographer, John Garraty (1953, p. 144), reports that he was unable to find a single instance after the publication of this research in which Lodge disparaged the Irish.

By the beginning of the second decade of the twentieth century the Irish politician was an unremarkable feature on the Boston scene. Few among the Brahmin class any longer regarded the Irish as interlopers or threats to Yankee "racial" survival. Recourse to the paternalism of *noblesse oblige* was a dead issue, and few chose to raise the question about the viability of democracy in a city of immigrants. Many who had once been hostile to the Irish were ready at last to place their resources of prestige and wealth at the disposal of Irish Catholic politicians.

Yankee acceptance is finally best illustrated by the pattern of Yankee endorsements in Boston mayoral elections. In four instances after the turn of the century the Good Government Association endorsed Irish Catholics (Thomas Kenny in 1914, John Murphy in 1921, and Frederick Mansfield in 1929 and 1933). In the first three of these elections the main opponent and eventual winner was James Curley. In 1933, however, Mansfield's unsuccessful opponent was Yankee Malcolm Nichols. A protégé of the Republican boss Charles Innes, Nichols had won only the most reluctant backing of the GGA in his race in 1925 *(NYT*, 29 Oct. 1925). In 1937 Boston's Yankee elite—Lowells, Storrows, Parkmans, and Shattucks—once again came to the aid of an Irish Catholic, this time Maurice Tobin, an old ally of James Curley now bent on challenging him (Lapomarda, 1970). If some few in the Brahmin community still regarded the Irish with bitterness,[12] it was nevertheless attended with resignation. On the whole, Yankee Boston had accepted the rule of the new majority.

STRATEGIC ADJUSTMENT
TO ETHNORACIAL TRANSITION

For Yankee Republicans the ultimate futility of attempting to wrest the mayoralty from Irish Democratic hands was apparent early in the transition period. Although head-to-head contestation across the party divide could not be abandoned, chances for success were clearly dependent on the periodic tendency of the factionalized Democratic party

[12]In 1920 Moorfield Storey complained in the Godkin lectures at Harvard about the dishonesty and inefficiency of "men of Irish descent . . . devoted to Irish interests and working to secure place and power for Irishmen [pp. 132–133]." And Gamaliel Bradford wrote in a letter in 1931 that "the Irish and Jews . . . seem to be the real Boston at present. . . . The Boston that we loved . . . seems forever departed [Brooks, 1934, p. 358]."

to disintegrate entirely. As early as 1886 the Republicans were driven to the desperate strategy of surreptitiously funding a third-party Labor candidate in the mayoral contest in the vain hope of splitting the Democratic vote (Nation, 1886). Reflecting on the decade before 1909, the year the new Boston city charter introduced nonpartisan ballots in local elections, Andrew Peters (1930), a Democratic reform mayor of Boston from 1918 to 1921, wrote: "For some years prior to the adoption of the new charter, the Republican party had been growing weaker in numbers; and it had become obvious that only a Democrat had any chance of being elected to a major office [p. 83]." Given this diminished possibility for a productive strategy of partisan contestation after the election of Hugh O'Brien, Yankee Republican energies were turned more and more toward multifaceted strategies of subversion and withdrawal.

In contrast, in the decade before the turn of the century, Yankee Democrats associated with the Mugwump movement pursued a vigorous strategy of cooperation with the Irish. These efforts, as we have seen, produced the mayoralties of Nathan Matthews and Josiah Quincy. William Jennings Bryan's nomination for the presidency in 1896 and the restiveness of Boston's younger Irish ward leaders, which found expression after the death of Maguire, combined to close off the opportunities for what the Yankees called "Independent Democracy." Although both Matthews and Quincy played elder party statesmen roles in the city Democratic caucus that nominated Patrick Collins for the mayoralty in 1901 (BET, 12 Nov. 1901) Yankee Mugwumps increasingly joined in or actually came to lead efforts broadly opposed to immigrant power in which Republican Yankees had already established a solid record of action.[13] The formation of the Good Government Association in Boston in 1903 and the rise of the Immigration Restriction League, among other developments, marked the consolidation of Yankee energies across party lines after 1900 (McFarland, 1975, pp. 84–85, 135, 175).

Subversive strategies—in this case, efforts to change the formal structures and constraints within which politics occurred to the disadvantage of the newly powerful group—were promoted through reform. Routinely enlisting the aid of the Massachusetts General Court (the state legislature), still thoroughly dominated by Yankee Republicans, reformers pressed for state assumption of municipal powers and other forms of intervention, charter reform, and metropolitan government in an effort to limit Irish Democracy. In 1885 and 1887 the Boston Reform Association

[13]Matthews, however, later affiliated himself with a Yankee reform organization established in 1909 called the Citizens' Municipal League. The League endorsed James Storrow, a Brahmin and national Democrat, against John F. Fitzgerald in the mayoral contest in that year (BET, 18 Nov. 1909).

sought and won the imposition of debt and tax limits on the city. Curiously, Hugh O'Brien himself joined in the call for such limits in order to establish a reputation for fiscal conservatism and win favor in Yankee quarters (Hanford, 1932, p. 91). Most subsequent mayors, however, including Hart and Peters, neither of whom had need to court Yankee tolerance, complained bitterly about the state-imposed constraints (Parkman, 1932, p. 132; Peters, 1930, p. 68). In 1885 the state legislature also placed the Boston police force under the control of a board appointed by the governor. Temperance advocates in the reform movement had urged such action on the grounds that the police were not enforcing the liquor license laws with sufficient vigor (Hanford, 1932, p. 91). Indeed, between 1885 and 1908 the Massachusetts legislature passed a total of 400 special laws dealing solely with the city of Boston (Gelfand, 1975, p. 8).

The anti-Irish implications of these activities were clear. As Andrew Peters (1930) put it delicately, "whether the rapid influx of a new racial element into the Boston electorate, with the consequent increase in Democratic voters, has influenced the actions of Massachusetts Legislatures, which are normally Republican, is a matter of pure conjecture [p. 74]." However, for Michael Hennessy, a Boston newspaper reporter and popular historian, there could be no doubt about the motives of the General Court. The Boston Charter Bill of 1909, Hennessy (1935) wrote, "was opposed by substantially a unanimous vote of the Democratic Party. It was designed to clip the wings of the then Democratic mayor of Boston, John F. Fitzgerald [p. 124]."

Fitzgerald was widely regarded as a different sort of Irishman from his predecessors in the mayoralty, Hugh O'Brien and Patrick Collins. O'Brien and Collins were gentlemanly figures; Fitzgerald was a ward boss from the North End, a Young Turk who had opposed Maguire's Yankee mayor strategy. When Fitzgerald ran for the mayoralty on Collins's death in 1905, the Brahmin business community, working through the Good Government Association, mobilized in opposition, backing Louis Frothingham, Republican speaker of the Massachusetts lower house. After Fitzgerald won handily, the GGA continued its campaign against him by calling on the state for an investigation of the city's financial affairs (Abrams, 1964, pp. 146–147). Seeking to head off state intervention, Fitzgerald created a local investigating commission. The commission, appointed by the mayor in 1907 and dominated by GGA members, recommended two alternative charters to the voters of Boston. One of these was ratified in 1909.

The new charter introduced nonpartisanship and at-large aldermanic elections, solid planks in the reform platform. Although the mayor's veto power was strengthened by the new charter and his term lengthened to 4 years, his appointments henceforth had to be approved by the state Civil Service Commission.[14] Finally, the new charter created a city finance commission to oversee the city's money matters. Its members were to be appointed by the governor. None of these reforms, as it turned out, forestalled the return of Fitzgerald to city hall in 1910 after a 2-year hiatus or the subsequent successes of James Curley, a consummate party politician.[15]

The municipal reformers not only sought to undercut Irish rule through an attack on structural elements of the political system that they believed facilitated the maintenance of machine power, such as partisan elections, ward-based representation, and mayoral patronage powers, but also to dilute Boston's Democratic electorate with suburban voters through metropolitanization. Middle class flight from the city, which coincided with the growth of the immigrant population, was already evident by the 1850s (Warner, 1962, p. 14). As the prospect for Irish rule in Boston began to take shape, interest in the extension of the city's boundaries to encompass the Yankee suburbs grew accordingly. A report by a committee in the Massachusetts legislature stated in 1863: "It is necessary to adopt the metropolitan principle in order to prevent the elements which are destructive of property and laws from keeping practical control of the city [Blodgett, 1966, p. 120]." Between 1867 and 1874 the city carried out five significant annexations, but growing suburban resistance brought the city's territorial growth to a virtual end. During the period of transition (1885-1930) only one annexation, that of Hyde Park in 1911, succeeded (Ward, 1963, pp. 139-140, 267).

In 1905, badly frightened by the prospect of Fitzgerald as mayor, the *Transcript* (18 Nov.) issued a plea for metropolitan government that sounded much like that in the 1863 legislative report. Yet it would be incorrect to interpret the interest in metropolitan government solely in terms of the desire to diminish the strength of the growing Irish electorate. Toward the end of the century motives for metropolitanization were as much a product of the belief in the desirability of growth and metropolitan tax equity as they were of a fear of Irish power (Beale, 1932, pp. 120-121). In 1895 a plan for a "Greater Boston" was referred to the next General Court session but no action was taken. Twenty years later Andrew Peters took up the cause of metropolitan government, mixing a subtle nativism with the themes of boosterism and efficiency in his argument. Not only would size attract business and services be rationalized, he argued, but the inclusion of the suburbs in a federated city

---

[14]This provision was rescinded in 1930. The city also returned to ward elections in 1924, only to revert to at-large elections in the 1949 charter reform.

[15]For a similar assessment, see Stoffer (1923).

would give the city a "better balanced citizenship." The commuters to
Boston are people of character and intelligence, he wrote, and it "should
be their duty . . . to take part in the general government [of the city]
[Beale, 1932, pp. 120–121]."[16] Once again nothing came of this proposal. By 1924, however, the issue had clearly lost its overtones of ethnic
and party conflict. In that year James Curley introduced a Greater Boston
Bill in the General Court (it failed), and in 1930, as mayor, he pushed yet
another proposal for metropolitan federation. The metropolitan strategy
as a device to undercut Irish power was thoroughly dead.

   Although contestation and the subversive strategies of reform and
metropolitan reorganization brought Yankee activists into direct or implicit confrontation with the Irish Democrats of Boston, other paths of
adjustment to the loss of political dominance were more oblique. Some
members of Boston's oldest families withdrew spiritually to the gratifications of a new Anglophilism (Mann, 1954, p. 7; Solomon, 1956, pp.
56–57). The period 1875–1895 was also a frenetic time of club-founding,
as Brahmin society sought to insulate itself from the unpleasantness of
the new order (Williams, 1970, p. 4). Still others turned their energies to
the cultural realm. "More and more," wrote Barrett Wendell in 1902, "it
seems to me that the future of our New England must depend on the
standards of culture which we maintain and preserve here. The College,
the Institute, the Library, the Orchestra . . . are the real bases of our
strength and our dignity in the years to come [Green, 1966, p. 23]."

   More significant strategic withdrawals occurred through the charities movement and the effort to restrict immigration. The former diverted
energies from politics to philanthropy as a means of social control: Up-
lift, as Huggins (1971, p. 191) has observed, was an attempt at management. The restrictionist movement represented a withdrawal from the
battle with the Irish in order to open a new front against a different foe.
A number of streams of Brahmin resistance were symbolically united by
the endorsement in 1901 of the Immigration Restriction League by the
Boston Associated Charities. The latter was headed by mainstay Yankee
Mugwumps, Charles Codman and Robert Treat Paine, although many
of the central figures in the social work movement, such as Robert Woods,
were not themselves Boston-bred Yankees. The Immigration Restriction
League had been founded by young men of the generation following the
Mugwump glory. After the turn of the century, however, the League
drew increasingly upon the strength of the Democratic Mugwumps,
until at its peak the membership reflected the full political heterogeneity
of Boston's Brahmin class.

---

[16]For a similar argument, see also Stoffer (1923, p. 83).

The effects of withdrawal were, as the term suggests, to create for a good many Yankee activists a life apart from the maelstrom of local politics. With the onset of World War I and the subsequent passage of immigration quotas during the 1920s, immigration diminished sharply. The League's work was done, although a small core of members continued to guard the gates through the period between the wars. By the 1920s the public energies of the Yankees in Boston were nearly dissipated. Nothing the Brahmins had tried had restored their place in the political life of the city or stemmed the Irish advance.

TOWARD AN EXPLANATION OF YANKEE ADJUSTMENT

American history has surely witnessed few ethnic divisions so filled with animosity, distrust, and disdain as the one between Yankee New Englanders and the Irish. For the better part of the nineteenth century these two groups regarded one another across a divide of class, political, and religious differences so broad that genuine accommodation seemed impossible. It is remarkable, then, particularly in light of the deep sense of fear that Yankees felt for the survival of their culture, "racial" stock, and position, that ethnic political transition in Boston was accomplished so peacefully. The explanation for the character of this transition, I would suggest, lies in a combination of factors that served gradually to diminish Yankee anxieties and resign them to their situation. Although it is difficult to determine the relative significance of these factors, I would argue nevertheless that their decisiveness may probably be laid to the tempering effects of the generalized context of democratic commitments acknowledged by the Brahmins as well as to the inescapable fact that Yankee voters composed a minority.

One factor that reassured a number of Yankee observers, such as Francis A. Walker and John D. Long, was evidence by the turn of the century of Irish social mobility. Hard work and ambition were values the Brahmins understood and respected. Census data from 1900 do indeed show clearly a declining concentration in menial occupations between first and second generation Irish workers with corresponding increases in skilled occupations (see Thernstrom, 1973, pp. 123–133; Hutchinson, 1956, p. 174). Upward social mobility among the Irish, however, was indicative for Brahmin writers of more than ambition. It suggested finally an assimilatory capacity on the part of the Irish, evidence of the ability to participate in what the Yankees were convinced was the true American experience. From Ralph Waldo Emerson's belief in the "smelting pot" to John D. Long's conviction that the American citizen of the twentieth century would emerge from an "amalgam of all nationalities," one strand in the Brahmin belief system was the expectation that the diverse immigration would eventually produce an American blend.

A second, though perhaps less important, factor that helps to explain the character of the response to the loss of political dominance in Boston was the maintenance of Yankee hegemony in Massachusetts state politics. Although it is not clear whether frustrated Boston Brahmins actually found outlets for their ambition in state government, the statehouse was nevertheless a Yankee preserve during virtually the entire period of transition. It is not unreasonable to suppose that such dominance provided a certain measure at least of psychological reassurance that some aspects of the Yankee world remained stable in a world of flux. Between 1884 and 1930 only one Irishman—David Walsh—succeeded in winning the governorship and that for only 2 years. In that 46-year span the Republicans elected 13 governors, who controlled the statehouse for a total of 37 years. Four Democratic governors accounted for only 9 years, of which 3 years could be attributed to the Bullmoose split in the ranks of the GOP. Three of the Democrats were themselves of Yankee stock, and a number of state Democratic leaders were avowedly anti-Catholic (Lockard, 1959, pp. 45, 119).[17] The Republican party organization at the state level, reputedly among the strongest in the nation, managed to control both houses of the Massachusetts General Court during the entire period between the Civil War and 1948. Under these circumstances, relinquishing the city was perhaps less crippling to Brahmin activists than it might have been.

Yet Boston was important to the old stock New Englanders, not only as the seat of their business enterprises but as the symbol—the hub—of their civilization. It was the genius of the Irish that they understood this. Thus Irish sensitivity to the implications of transition for the Yankees constitutes a third factor that influenced the nature of Yankee adjustment. O'Brien's "reform" mayoralty and the choice of Patrick Collins, with his Harvard law degree and diplomatic experience in England, did much to assuage Yankee fears. Irish politicians were also consistently solicitous toward business: Fitzgerald, for example, coined the slogan, "A Bigger, Better, and Busier Boston," as he promoted port development (Shannon, 1963, p. 209).[18] In addition they were not unwilling to share patronage tidbits with selected Yankees. Robert Grant (1934), a Mugwump lawyer, described himself as "a grateful but astonished officeholder" upon his appointment by O'Brien as a Water Commissioner in

[17]See also Hennessy (1935) and Huthmacher (1969, pp. 14–18).

[18]In 1890 the Irish Congressman from Boston, Joseph O'Neil, announced, "This is a business age, and we are essentially a business community [Blodgett, 1966, p. 145]."

1888. "I was at a loss," he wrote in his autobiography, perhaps naively, "to know why the Mayor should have consented to nominate one who . . . could be of no assistance politically [p. 172]." Patrick Maguire played a crucial role in this and other efforts to distribute patronage spoils to others besides claimants of Irish ancestry (Blodgett, 1966, p. 60).

Other factors also undoubtedly help to account for the relative ease with which Boston Yankees came to accept Irish power. The transfer of anxieties from the Irish to the newer immigrants clearly served to diminish the fearsomeness of the Irish threat. The fragmented character of the Yankee response to Irish power was yet another factor that helped to resign the Brahmins to their loss. This fragmentation arose from the pluralism of the Yankee community with its Mugwumps and Republicans, its different generations, its social workers and merchants, its restrictionists and cultural pluralists. Some cooperated with the Irish; others left the city. Some pursued reform; others philanthropy. None of these strategies stemmed the Irish advance. It seems probable that the relative ineffectiveness of a fragmented response sapped the spirit and energy of those who sought to resist. A united response might in contrast have given hope, prolonging and intensifying the struggle.

By the 1930s Yankee observers saw that 50 years of Irish dominance had not produced radical changes in the conduct of municipal affairs. If the style and sympathies of the Irish differed from those of the Brahmins, a low tax rate had nevertheless become a universally important symbol. For Yankee Boston there were no longer any powerful reasons not to accept the well-established fact of Irish rule.

SOME LESSONS: BENCHMARKS IN THE PROCESS OF ETHNORACIAL TRANSITION

From the vantage point of contemporary times, from which many of the nasty details of daily politics in turn-of-the-century Boston cannot be seen and do not obscure the issue, the process of ethnic political transition taken as a whole reveals a political system with many adaptive and comparatively benign characteristics. This is not to say that the political system of the city was open to all comers, nor necessarily that it was responsive or honest.[19] But it did manage to accommodate intense ethnic conflict, interwoven with elements of class and religious struggle, in a peaceful and nondestructive way.

[19]Thernstrom (1973, p. 132) points out, for example, that the task of producing group economic benefits from the political control the Irish enjoyed was extremely slow and difficult.

Was Boston unique? Was the contest between the Yankees and the Irish *sui generis?* We may certainly draw some conclusions from this particular case that serve not only to summarize matters but to provide benchmarks to watch for as we investigate the contemporary transition process in Atlanta and Detroit.

1. The process of transition is not likely to occur overnight. In Boston it surely could not be regarded as complete for several decades after 1884, for Yankee opportunities for political victory were not entirely closed off until the 1930s. The new pattern of political power is slow to take form not only because the new group is still consolidating its power, but because the displaced group is in no sense subject to a purge or harassment. Although the fortunes of the displaced decline, they still control significant resources that permit the establishment of political opposition. The new group not only allows this, but it also apparently seeks to tap those resources for its own purpose.

2. During the process of transition, the newly emergent group is likely to suffer temporary electoral reverses, but the weight of its voting bloc will reestablish its dominance. When factional splits within a newly powerful group coincide with or stimulate the consolidation of opposition resources, the displaced group is likely to win a temporary period in office, as the Yankees managed to do several times. The new group, realizing the costs of division, will smooth over its differences in order to regain political dominance. Eventually, factions within the new group will come entirely to structure electoral competition, as they did in Boston.

3. By the time transition begins, competition between the two ethnic or racial groups is likely to be pursued through peaceful means. The moment from which we may date the beginning of transition is the capture of the mayoralty by a candidate from an ethnic or racial group that heretofore occupied a subordinate place in the city's social and political structure. But such a victory marks the onset of transition only when the newly emergent group constitutes a majority or near majority of the city's population, thus making future control of the mayoralty by candidates from the new group highly likely.

Although violent conflict between politically established and emerging ethnic groups has been common in American history, it seems to occur principally in the earliest stages of the new group's mobilization, long before it is ready to challenge for top political office. There is evidence that suggests indeed that such violence may be more a response to the new group's efforts to break into the labor market than a reaction to its political stirrings (Eisinger, 1976). Mob violence against the Irish, for example, began in the mid-1830s and ended for all practical purposes 20 years later, before the Irish had achieved even the most minimal level of political visibility. Collective violence directed at blacks in cities occurred mainly during and immediately after the two world wars, when blacks were making great advances in the labor market.

By the time the new group has achieved a local electoral majority, however, the group's claims to power appear to be evaluated in light of its voting strength. Commitments on the part of the previously established (now displaced) group to the principles of majority rule and to the sanctity of elections apparently preclude the resort to violent means to change the turnabout in the patterns of formal power.

4. A major factor contributing to the peaceful character of transition seems to be the tendency of the newly victorious group to make overtures and concessions to the displaced group. To the extent that a winner is concerned with the sensibilities of the losers, it is out of a mix of motives. On the one hand the winner knows that today's victor may be tomorrow's loser. But more important, the new victor often wishes to tap the resources controlled by the displaced group—resources of money, experience, expertise, and prestige in particular—in order to govern with maximum effectiveness. Thus the newly victorious group is likely to seek to make transition palatable to the displaced, for example, by appointing people from that group to office and by offering a variety of other, often symbolic, assurances (such as the firm commitment to business development promised by the Irish mayors).

5. Strategies pursued by members of the displaced group that are designed to deal with the loss of formal political power are likely to be diverse and uncoordinated. Although the natural pluralism of the displaced ethnic or racial community may be set aside in a highly visible electoral contest with a candidate of the newly emergent group, those divisions tend to reassert themselves and influence strategic adjustments between elections. Different elements in the community, whose only common attribute may be race or ethnicity, tend to have differing estimates of the degree of threat posed to them by the new order. People divided by ideology, age, and class are likely to find that they cannot agree on appropriate strategic responses. The diversity of responses lessens the chances of reestablishing political dominance.

6. Among displaced elites the public expression of overtly racist attitudes or the outright attempt to exploit racial fears as strategies of adjustment seem quickly to become unacceptable. Overt racism—the expressed belief in racial or ethnic inferiority—loses its appeal for a group that has lately lost its position of political dominance. Racist pronouncements are likely to be seen both as unworthy and as politically unwise for a new minority group. Displaced elites may instead criticize the new regime by judging it against a set of neutral-sounding standards.

However, these standards—such as nonpartisanship, sound administration, efficiency, and experience—often seem to serve as surrogates for more venal attitudes and thus assume the currency of code words. These seem to function for many as a way of expressing concern in a publicly acceptable way, masking more basic fears.

7. As transition progresses, displaced elites may discover a new threat or problem, the contemplation of which may divert their attention from the dilemmas of displacement. The Yankee concern over the "new" immigration illustrates the dynamics of this diversion. Such transfers of concern may not only provide some measure of psychological relief but may also serve on occasion to unite the displaced group and the new victors in a common cause. The diversion of attention to a new problem may also constitute a tacit way of signalling acceptance of the new political situation.

8. Elites among the displaced group are neither likely to abandon the city in wholesale numbers nor to withdraw entirely their resources from the city and the new governors. In Boston Irish politicians sought successfully to use Yankee resources. Many among the Yankee elite remained in the city as transition progressed: some to cooperate with the new governors, others to fight them. Many stayed to foster business enterprises and the business climate of the city, or to maintain the cultural and philanthropic institutions of the city.

SUMMARY

In Boston the process of ethnoracial political transition resulted in the full transfer of political power from the hands of one ethnic group, favored by virtue of class status and heritage, to another, situated initially at the bottom of the white status hierarchy. There is no evidence to suggest that the Irish were the pawns, unwitting or otherwise, of Brahmin wirepullers. Society was transformed in a way that could not perhaps have been predicted from the nature of Irish–Yankee relations prior to 1880. A combination of commitment to democratic norms and a pragmatic electoral calculus formed the groove along which Yankee adjustment proceeded. The question that remains is whether the character of transition in Boston was a product of New England genius or whether it illustrates a more general trait of adaptability in the American political system.

3

Transition to Black Rule in Modern Detroit and Atlanta

Certain of the contrasts between Detroit and Atlanta could not be sharper. Detroit is a great sprawling manufacturing center, self-styled "capital" of the industrial crescent that hugs the southwestern shores of the eastern Great Lakes. The fifth largest city in the nation, with 1.3 million people, Detroit is the home of ethnic groups from places as exotic as the ancient Chaldean region of Iraq and as familiar to the American experience as Poland and Italy. It is a blue-collar union city of tough sensibilities, kin to Chicago and Cleveland, which flank it on either side. It is, of course, a northern city, cold in the winter, gray and gritty in its general aspect.

Atlanta, too, is a capital, not just of the state of Georgia but of the southeast. Less than one-half million in population, it is a city nourished by dreams of its own gentility and energy. The name Peachtree, the city's major thoroughfare, evokes all those peculiarly Southern associations of plantations, society, and warmth, which, if stereotypical, are not wholly irrelevant to modern Atlanta and its sense of itself. It is a young city in contrast to Detroit, having emerged only in the post-Civil War period as a regional depot. It is an economically healthy city, whereas Detroit lives on the edge of financial collapse. Atlanta's residents are white and black, but not ethnic. There is no union movement to speak of, for the city's lifeblood is in distribution, warehousing, and white-collar office occupations, rather than industry; besides, it is a southern city, traditionally hostile to organized labor.

However, there are important similarities between the two cities, as well. For instance, both have spectacular new downtowns developed in the 1960s and 1970s: Indeed, the flamboyant centerpieces of both places were designed by the same architect, John Portman. Both cities have long traditions of reform in politics, and in the early 1970s undertook charter revisions that strengthened the powers of their mayors at the expense of the city councils. It is no small irony that the first beneficiaries of these changes, developed largely by white reform commissions, were two black mayors. In fact, the most striking similarity is perhaps that both cities now have black majorities and black mayors.

Detroit and Atlanta also share a history of racial tension. In both places the problem of race relations has run like a thick rope through the finer weave of politics: In many ways race has been and still is the overriding issue in local politics and government. This chapter explores the problem of race in the modern politics of the two cities for the purpose of contrasting the extraordinary tensions of the quarter century after 1945 with what we shall later see to be the comparatively easy adjustment whites in both places have made to black rule in the mid-1970s.

THE DEMOGRAPHICS OF RACIAL TENSION

Detroit and Atlanta have both been boomtowns in their time. In the first half of this century Detroit's auto plants were perhaps the most compelling industrial magnet in the nation, quadrupling the city's population between 1910 and 1950 (see Table 3.1). In the 1970s motor vehicle manufacture still accounted for more than one-quarter of a million jobs of the nearly 1.6 million total employed in the Detroit metropolitan area, by far the single largest source of work. The General Motors Corporation and the Chrysler Corporation both currently maintain several major plants within the city; Ford facilities, absent now from the city itself, are scattered throughout the metropolitan area. Names of the fathers of the automobile industry adorn many of Detroit's streets and expressways. There is still no better vivid symbol of the city than the massive Diego Rivera murals at the Detroit Art Institute depicting the human toll and promise of the auto assembly lines.

The first great growth spurt in population occurred in the years

TABLE 3.1
Black Population Growth in Atlanta and Detroit

Year	Atlanta			Detroit		
	Total population (in thousands)	Total black	Percentage black	Total population (in thousands)	Total black	Percentage black
1910	154	51	33	465	5	1
1920	200	62	31	993	40	4
1930	270	90	33	1,568	120	7
1940	302	104	35	1,623	149	9
1950	331	121	37	1,849	303	16
1960	487	186	38	1,670	487	29
1970	496	255	51	1,511	660	44
1975 (est.)	470	265	56	1,385	770	56

Source: U.S. Bureau of the Census.

around World War I, a product in part of a public relations gambit by the Ford Motor Company. In an effort to expand and upgrade its workforce and to reduce turnover, the company announced in 1914 the advent of the 8-hour, $5 day, abolishing in the process the wage differential between skilled and unskilled workers. This unprecedented wage—"profit-sharing," Ford called it—drew thousands of people to the city (Conot, 1974, p. 164). By the end of World War I the population had more than doubled that of 1910.

Drawing principally on the European immigrant labor pool until the 1920s and then, increasingly, on the swelling black migration from the South, the city reached a high point in its growth by the census of 1950. The expansion of Detroit's black population had been steady but slow prior to 1940. However, with the increase of skilled employment opportunities for blacks brought about by the labor requirements of war mobilization (Weaver, 1946, p. 78), the city's black population doubled between 1940 and 1950. Then, set against a marked increase in postwar white suburbanization, the black proportion in the city grew from 16% in 1950 to a clear majority in 1973.

As Detroit began to lose population in the 1950s, Atlanta began to surge. Atlanta had by this time achieved modest prominence in a successful competition for regional dominance. Founded as a railroad center, the town had long fulfilled a depot function. The establishment of air routes in the late 1930s and the relocation in 1941 of the fledgling Delta Airlines from Monroe, Louisiana, to Atlanta made the city a central air transfer point for the entire Southeast. Today its residents boast that Hartsfield International Airport is the second busiest in the nation.

The city's key role in the regional transportation network made it a major distribution and service center, enabling it to eclipse its nearby rivals, Birmingham, Nashville, and Tampa. In 1972, 28% of the city's workforce was engaged in wholesale and retail activities, a greater proportion in these occupations than in any of the top 20 employment centers in the nation (Hartshorn et al., 1976, pp. 1, 5).

Atlanta has experienced two periods of significant growth in this century. One was associated with the first "Forward Atlanta" program of the Atlanta Chamber of Commerce. Launched in 1925, this million dollar promotional scheme to enlarge the city's industrial base was a model of successful boosterism. Industries attracted to the city added nearly $30 million in annual wages to the total city income between 1926 and 1929 (Allen, Jr., 1971, p. 149). Population grew in the 1920s by 35%. The second major growth period occurred during the two decades after 1950. A successful annexation in 1952 added approximately 100,000 people to the city, but Atlanta's natural growth can be laid mainly to the production of jobs in the 1960s. Partly as result of a second "Forward Atlanta" campaign,[1] the Atlanta metropolitan area enjoyed an annual average increase of 23,000 jobs between 1960 and 1972. In that decade the skyline of the city was transformed, as corporations built regional and even national headquarters there. The city's population reached a peak of just under one-half million around 1970.

Black population growth, as a proportion of the total, scarcely changed in the 50 years between 1910 and 1960. It is true that the annexation of 1952 stemmed a more rapid increase, but the decennial percentages are nevertheless remarkably stable. The significant change came during the 1960s, at the end of which blacks had achieved a bare majority. The growth rate of the black population of both cities in the 1960s (36% in Detroit and 31% in Atlanta) is consistent with the average of 35% among central cities in the 66 largest metropolitan areas of the nation ("Where Blacks Are Moving," 1971, p. 24).

Rapid growth of a minority population after a long period of slow growth, combined with the relatively sudden achievement of majority status, may be expected to induce certain social strains. Indeed, both cities suffered civil disorders in the 1960s, although Atlanta's minor troubles in 1966 (and one incident in 1970) pale in comparison with the Detroit upheaval of 1967. But there were other demographic factors besides the rapid growth of the black population during a generally tense decade that contributed to a wide sense of racial unease in the two

[1] This $1.6 million promotional effort was suggested by Ivan Allen, Jr., in 1961 when he was president of the Atlanta Chamber of Commerce. He rode the program to the mayoralty in that year.

cities—for example, the extremely high level of residential segregation in both places. According to an analysis by Taeuber and his associates of census tracts in 109 American cities, residential segregation in Atlanta has actually increased since 1940 (Sørensen et al., 1975), due in part to urban renewal policies pursued during the 1960s that destroyed much low-income housing and reconcentrated the displaced population (Stone, 1976, p. 117). In 1970 Atlanta ranked nineteenth among the 109 cities examined by Taeuber, with a segregation index of 91.9.[2] The mean for all these cities in 1970 was 81.6. In Detroit residential segregation has declined slightly in each decade since 1940, lodging just above the mean in 1970 at 82.1. Expansion of black residential areas was carefully controlled in both cities through the use of urban renewal condemnations and freeway construction.

Neither Detroit nor Atlanta has been unique in their levels of segregation. Nevertheless, the fact that neither city stands out in the degree to which its respective black populations live in all-black neighborhoods scarcely mitigates the effects of residential segregation. Residential segregation is pernicious, but not simply for the humiliation a ghettoized people must bear. As David Harvey (1973) has pointed out, the spatial ecology of the city affords various population groups differential access to employment opportunities and public goods (for example, transportation, public health care, and recreation facilities), and it imposes widely varying externality costs (pollution effects, for example).

Other population characteristics familiar even to the most casual observers of American cities also suggest a basis for racial tension. In both cities, for example, as in the nation at large, black unemployment rates were generally double those of whites throughout the 1960s. The gap between per capita median black and white incomes in Atlanta in 1969 was exceeded by only 1 of the 25 largest cities in the nation, and in only 5 of those cities was black income actually lower. Although black income in Detroit was higher than in all the other cities in the top 25, except New York and Chicago, white income was relatively even higher. The black–white income ratio in Detroit was the eleventh largest within this group of cities.

What is particularly important about these unemployment and income data is that both cities are popularly thought to be meccas of opportunity for blacks (Range, 1974). Interview respondents in both cities stressed the undeniable size and prosperity of the local black mid-

---

[2]The segregation index may be read as a percentage of one racial group that would have to change residence to blocks occupied by members of the other race in order for the city to exhibit an unsegregated or proportionate residential distribution.

___ class. Furthermore, both cities ranked high in 1970 in a "quality of life" study on an aggregate measure of economic health, central to which were measures of employment opportunities and economic achievement (Liu, 1976). The Atlanta Standard Metropolitan Statistical Area (SMSA) was classified in the Liu study in the "outstanding" category on the economic component scale, and the Detroit area was among those termed "excellent" (p. 95).

Clearly, however, there were in the 1960s (and still assuredly are) substantial levels of poverty and deprivation among blacks in the two cities. A highly suggestive finding that not only corroborates the continued existence of a deprived population but also offers a glimpse of the character of the social strains that might be generated as a result emerges from the same quality of life study. Despite their high aggregate levels of economic health—measured not only by income and wealth but by consumption patterns, employment, economic diversification, and productivity—both cities ranked low on a measure called the "social component." This is a summary factor measuring, among other things, "the level and potentiality of the development and flourishing of individual independence and dignity [and] the differences between the actual and desired levels of equality or justice in seeking employment and housing, in commanding goods and services, etc. as a result of race, sex, and spatial discrimination [Liu, 1976, pp. 128–129]." It is, in short, a measure of relative deprivation. Whereas the Atlanta metropolitan area ranked seventh out of 65 SMSAs on the economic measure, it ranked only 44th on the social component. The Detroit area, ranked 28th on the economic measure, was next to last among the 65 SMSAs on the social measure (p. 130). To the extent that poverty and deprivation have been at the root of black political militancy, it would certainly have been reasonable to expect that whites in the two cities would have anticipated and sought to deal with black rule with deep misgivings, profound wariness, reticence, and, above all, great defensiveness.

RACE AND POLITICS IN DETROIT AND ATLANTA

Racial antagonism was a fixture of daily political life in Detroit and Atlanta in the years immediately preceding the achievement of black rule. In both cities the black community had traditionally occupied places in large local voting coalitions that crossed racial lines, but here the similarity ended between the respective political roles of the two black populations.

After World War II Detroit blacks increasingly figured as part of a frequently shaky, often nominal, coalition in local mayoral contests

that was made up, additionally, of the UAW, the CIO, and white liberals. The record of this voting bloc was a discouraging one, however:
Between the end of the war and the election of Coleman Young in 1973,
blacks were on the winning side with only one mayor, Jerome Cavanagh
(see Table 3.2). Even at other levels of local government blacks were
slow to realize gains. At-large aldermanic elections, for example, prevented the city's black minority from breaching city council chambers
until the victory in 1957 of William Patrick, who served until 1965 as the
lone black councilman on a nine-person council in a city one-third black.
In the 1965 elections Patrick declined to run but was replaced by another
black, Nicholas Hood, who served alone until a special election in 1968
brought a second black to the council. In 1969 these two men were
joined by a third black, Ernest Browne, in a city now nearly half black.
Although Cavanagh had appointed Arthur Pelham, a distinguished black
economist, to head the city controller's office in 1961 (the first black to
occupy a major policymaking position in Detroit government), at the
time the 1967 riot occurred, the only major black official on the local
scene was an aide to the mayor (*Report of the NACCD*, 1968, p. 185).

A brief survey of Detroit mayoral politics prior to Coleman Young's
election in 1973 suggests not only the pervasiveness of racial issues but
the persistent failure of black influence. Detroit's majority white coalition yielded only infrequently, and then most reluctantly, to black demands for more adequate representation. After the only major success
for black voters—namely the mayoralty of Jerome Cavanagh—white voters returned city hall to a candidate thought to represent their racial
interests more faithfully.

The city was governed through most of the 1950s by Albert Cobo, an
avowed Republican conservative. His initial victory in 1949 was achieved
at the expense of a liberal Democrat named George Edwards. Conot
(1974, p. 403), among others, has interpreted that contest as a racial
confrontation. Edwards took a position in favor of open housing and

TABLE 3.2
Postwar Detroit and Atlanta Mayors

Detroit		Atlanta	
Edward Jeffries	1939–1947	William Hartsfield	1936–1940
Eugene Van Antwerp	1947–1949	Roy LeCraw	1940–1942
Albert Cobo	1949–1957	William Hartsfield	1942–1961
Louis Miriani	1957–1961	Ivan Allen, Jr.	1961–1969
Jerome Cavanagh	1961–1969	Sam Massell	1969–1973
Roman Gribbs	1969–1973	Maynard Jackson	1973–
Coleman Young	1973–		

increased public housing. Cobo played openly on the fears of white ethnic homeowners, who broke with the liberal UAW leadership to defeat Edwards. Once in office Cobo's first act was to halt existing plans for expanded public housing. His mayoralty is associated with a series of policies in housing and freeway construction designed literally to contain the city's growing black population.

On his death in 1957 Cobo was succeeded by Louis Miriani, Detroit's "Little Flower." Miriani served 2 months as acting mayor before he was elected to his own 4-year term with broad backing from labor, business, and the newspapers. Miriani's mayoralty was marked by an increase in police harassment of blacks, part of the mayor's war on "Negro crime." When he was challenged in 1961 by political newcomer Jerry Cavanagh, a 33-year-old liberal in the Kennedy mold, Miriani fought back by charging that Cavanagh was a mere candidate of the Negroes. With the UAW, the Chamber of Commerce, auto companies, and the press behind him, Miriani did not expect to lose. Nevertheless, a heavy black turnout, 85% of which went for Cavanagh, and the support of white liberals enabled the young challenger to win. His victory marked the first time black voters were on the winning side in Detroit local politics in the postwar period. One of Cavanagh's first acts was to appoint George Edwards, the man who had lost to Albert Cobo in 1949, to the post of police commissioner.

Cavanagh's two terms in office offer a study in contradictions. From the beginning the mayor was convinced that the future of the city lay in the social programs of Washington's New Frontier and Great Society agendas. Particularly during Lyndon Johnson's presidency, Cavanagh became a national prototype of the new mayor who would lead his city out of the wilderness to regeneration in the land of cooperative federalism. Cavanagh was the only elected official chosen by the President to serve on the Task Force on Urban and Metropolitan Development. In 1966 the mayor simultaneously held the presidencies of the U.S. Conference of Mayors and the National League of Cities. Federal antipoverty money came pouring into the city, urban renewal was changing the face of a large residential area adjacent to the downtown, and private office construction boomed in the central business district for the first time in 25 years. Cavanagh won reelection in 1965 without serious opposition.

Nevertheless, the quality of black life in the city remained poor, seemingly impervious to the activity of the young mayor. Ghetto unrest increased in this period, culminating in the 5-day riot of August 1967. Forty-three people were killed, most of them at the hands of the police and National Guard. Damage ran to between $30 and $50 million. A number of major corporations seriously considered leaving the city in the aftermath.

The Great Society programs, which President Johnson had introduced to the nation in 1964 in a speech at the University of Michigan just
40 miles from Detroit, seemed strangely unresponsive to black needs. In
Detroit, the Kerner Commission concluded, the new programs simply
produced "a deepseated hostility toward the institutions of government
[*Report of the NAACD*, 1968, p. 286]." Police behavior toward ghetto
blacks remained an issue that aroused the deepest resentments all through
the 1960s. A National Urban League study conducted among Detroit
blacks in 1967 found that 82% believed that the police used unnecessary
force in their dealings with blacks (p. 302).[3] After the 1967 riot the city
was painfully slow to rebuild its devastated ghetto neighborhoods. Much
of the area destroyed in the riot was still barren in 1977, although work
by a black real estate developer was beginning on a new apartment
complex near the heart of the riot area.

In the workplace as well as on the streets there was evidence of
mounting racial tension and bitterness: In the late 1960s the emergence
of several black "revolutionary union movements" in the auto plants
signalled a growing gap between a large number of young black workers
and the established UAW leadership (Georgakas and Surkin, 1975). But
next to the riot, the most dramatic evidence of the quality of black life in
the city was the surging murder rate. During Cavanagh's first term the
number of murders—most of them perpetrated by blacks against other
blacks—averaged about 130 per year. After 1965, however, the numbers
increased to 214 in 1966, 281 in 1967, 389 in 1968, and 439 in 1969,
Cavanagh's last year in office (Aberbach and Walker, 1973, p. 21). The
crime situation provided fertile ground for racial demagoguery. From
1965 on Cavanagh came under attack from Councilwoman Mary Beck for
his "failure" to control black crime. Beck parlayed the "law and order"
issue into a mayoral candidacy in 1969, but she failed to make the runoff.
Cavanagh's mayoralty, full of promise at the beginning of the decade,
ended dismally. It was summarized aptly by a city councilman (I-137)
in 1976: "Cavanagh was a man of the 1960s, a rallying guy, a man for
causes. But he wasn't very good at keeping house."

By the middle years of the 1960s black voters in Detroit had managed to make several successful, although extremely tentative, challenges
to the white monopoly in local government. Yet despite the acquisition
of seats on the city council and school board, black people seemed in
certain essential ways almost as firmly locked into their traditional political role as they had ever been in the postwar period. Basic political
subordination and marginality in relation to the whites who controlled

---

[3]Joel Aberbach and Jack Walker (1973, p. 53), however, report somewhat lower figures
in their 1969–1971 study of Detroit.

city government were still the rule. Among whites—and even among
many blacks—the prevailing structure of power was taken to be the
framework in which solutions to the "racial problem" would be worked
out in the future. To the degree that local action could have any effect on
the black condition, what was done, it was assumed in most quarters,
would have to be a consequence of progressive white initiatives or white
responses to black supplications. The alternatives of genuine black power
or black rule and their possible implications, if considered at all, resided
in the vague future. Toward the end of the decade, the races in Detroit
regarded one another with open fear and distrust. In this context it
would have seemed groundless to predict that imminent transition to
black majority rule in that city would be met with white elite acquiescence.

   By 1969 Detroit blacks were ready to challenge the white monopoly
over city hall with the candidacy of Wayne County auditor Richard
Austin. His opponent in the runoff was Roman Gribbs, a Polish-American
and sheriff of Wayne County. A *Detroit Free Press* survey of voters concluded that "race and race-related topics—mainly the schools and the
police—form the overriding issue in the campaign [Widick, 1972, p.
207]." The two men nevertheless conducted a restrained campaign. "I
would rather have lost than heighten racial tensions," Gribbs later observed in his interview. "Anyway, if you'd won and you'd alienated
45% of the city, how could you govern?" The UAW endorsed Austin,
but the black candidate was unable to make sufficient inroads in the
white vote. White rank-and-file union members broke once again with
their liberal leaders to help defeat Austin by slightly more than 7000
votes out of one-half million cast.

   Gribbs has pictured himself as a "transitional" mayor, helping to
prepare the city for black rule. His pattern of appointments certainly
reflected a sensitivity to growing black power in the city: Approximately
40% of Gribbs's department heads as well as his deputy mayor were
black. "In my first year I made more black appointments than Cavanagh
had in his eight years," Gribbs claimed in the interview mentioned
earlier. Yet Gribbs also presided over the formation in 1971 of the police department's tactical STRESS unit (Stop the Robberies, Enjoy Safe Streets),
which became known as a killer squad in Detroit's ghettos.

   As the decade of the 1960s closed and the 1970s began, the mood of
Detroit blacks was bitter. Greenberg's 1969 study (1974, p. 59) of black
opinion in the East Side ghetto area revealed that 71% of his sample
exhibited a "high" receptivity to violence: One-fifth actually expressed a
willingness to take part in another riot, and two-thirds believed that the
1967 riot had been helpful to blacks. A survey done in 1971 found that

54% of black Detroiters still believed that blacks had gained from the riot (Aberbach and Walker, 1973, pp. 59–60). Widick (1972, p. 207) observes that, not surprisingly, many whites were jubilant in 1969 that blacks were turned back in their attempts to "take over." The prognosis for race relations in those years was hardly bright. Aberbach and Walker concluded their study of the city by noting "the evident distrust and hostility existing between the races" as the two groups struggled "to develop a satisfactory basis for social and political interaction," and that an "end to racial conflict or tension there is unlikely in the near future [pp. 61–62, 216–217]."

If the role of Detroit blacks was to play the frustrated partner in a losing coalition, that of Atlanta blacks was to serve as a silent voting bloc in the city's ruling coalition. Allied with white business interests and the press, this coalition stood throughout the postwar years in successful opposition to lower-class whites. Once blacks were allowed to vote in the mid-1940s, they did so through the following decade with "greater unity of purpose and in larger percentages than either lower or upper income whites [Greenberg, 1974, pp. 60–61]." White political figures—at least those in power—were fond of saying in those days that Atlanta government rested on a three-legged stool: business, the press, and black votes.

The architect of this coalition—and the man who coined the slogan that Atlanta was "too busy to hate"—was Mayor William Hartsfield, first elected in 1936. When blacks became a modest political force in the 1940s, Hartsfield recognized the utility of the black vote, particularly as a means of safeguarding the ability of the business community to pursue its "progressive" policies—that is, economic development—and to maintain overall control of the city. Hartsfield thus established a working relationship with a respected leader in the black community, attorney A. T. Walden, founder in 1949 of the black Atlanta Voters League. What black Atlantans accomplished by their support of Hartsfield was to maintain in office what was (for the South) a racial moderate. The mayor was never seriously challenged in his nearly quarter century in city hall, although in 1957 he had to face Lester Maddox, the restaurateur and arch-segregationist, making his first run for political office.[4]

Aside from assurances that the city's influential whites would not countenance a blatant racist in the mayor's seat, Atlanta blacks got little else for their participation in the winning coalition. Not until 1965 did

---

[4]Hartsfield did lose the mayoralty briefly (by only 83 votes) to Roy LeCraw in a contest that had nothing to do with racial issues. LeCraw served only 15 months in city hall before being called up to active military duty. Hartsfield returned to office in a special election in 1942.

blacks elect their first alderman to the city council. And in all the years of Hartsfield's tenure, the norm that governed appointments to boards and commissions was to maximize "respectability" rather than "representativeness," a rule that effectively barred blacks from such bodies (Stone, 1976, pp. 29–30).

The power of the Atlanta Voters League began to crumble slightly in the early 1960s as the sit-in movement split the black community on the issue of tactics (Greenberg, 1974, pp. 61–62). Ivan Allen, Jr., president of the Atlanta Chamber of Commerce and key mediator between black protestors and downtown merchants in the sit-in disputes, emerged as Hartsfield's successor in 1961. Allen continued the Atlanta tradition of racial moderation, a stance dictated in part by his reliance on black votes. His opponent in 1961 was Lester Maddox, and the issue in the campaign was race. Allen had wanted to base his run for city hall on his economic development program, but Maddox's challenge changed the nature of the debate: "I could promise all I wanted to about Atlanta's bright, booming economic future, but none of it would come about if Atlanta failed to cope with the racial issue. . . . Undoubtedly, [Maddox] would scream 'nigger-nigger-nigger' throughout the campaign [1971, p. 53]." Allen attacked Maddox, claiming that the restaurant owner would "bring another Little Rock to Atlanta." The strategy worked: Maddox lost badly, taking only 35% of the vote. Allen won all but 237 votes of the nearly 22,000 cast in the black precincts (p. 60).

In 1965 Allen won reelection without a runoff. Blacks supported him overwhelmingly again, in part in recognition of his testimony in favor of the 1964 Civil Rights Act. To remind black voters of his role in the hearings, Allen carried his transcribed remarks before the congressional committee with him in a leather-bound volume whenever he campaigned before black audiences.

Although one black was elected to the city council in 1965 (1 of 18 councilmen), black participation in Atlanta government through the 1960s was largely by proxy. The extent to which blacks were essentially powerless in this period is amply illustrated by Clarence Stone's study (1976) of urban renewal in Atlanta in the Hartsfield–Allen years. The book demonstrates how little of importance blacks actually got for their participation in the "progressive" coalition that ruled the city. In his first years in city hall Allen showed little interest in Atlanta's extensive low-income neighborhoods, although in 1965 an estimated 160,000 city residents lived in substandard housing (p. 117). Allen's development policies, however, focused on the central business district. After minor civil disturbances in 1966, in which the absence of neighborhood renewal plans was an issue, Allen did appear to embrace plans to refurbish the city's ghetto areas. This interest was neither fruitful nor sustained, however: "During his final years in office, Allen made it clear that he regarded economic prosperity as a more fundamental community need than relieving poverty or improving neighborhoods [p. 130]."

The election of 1969 marked the end of the reign of the Chamber-dominated coalition. When Allen (1971, p. 220) decided not to run again, he had quietly passed word of his decision to the city's white "top leaders" so they could begin a search for his successor. The eventual choice of the white business community was Rodney Cook, a conservative Republican insurance executive with a record of state and local public service. When black leaders were told of the choice, they refused to support Cook, and the coalition of the previous decades dissolved.

Although a black educator named Horace Tate made a tentative run for the mayoralty, Cook's most serious challenger was the city's vice mayor, Sam Massell. The vice mayor, a Jew, was regarded by the Chamber group as a political outsider and a maverick. Massell established himself as a friend of the black community from the beginning, promising among other things, to increase black representation in city government. He won solid black support. In the runoff Massell took 55% of the total vote, but only one-quarter of the white electorate supported him. In the same election, four blacks were elected to the city council, and a young man named Maynard Jackson was elected to succeed Massell as vice mayor.

THE BLACK MAYOR ELECTIONS

In 1973 neither Maynard Jackson nor Coleman Young was the first choice of the forces of opposition in their respective cities. In Atlanta Jackson was regarded by the established black elites as a man who had not waited his turn. Many of the men in the black community who would be expected to finance a black mayoral campaign preferred black State Senator Leroy Johnson (Range, 1974). In Detroit the Wayne County UAW believed that a white sociology professor named Mel Ravitz, then finishing a term as president of the city council, was the strongest candidate the labor–liberal–black coalition could field (I-19, 119).

It came as a general surprise, then, when both Jackson and Young demonstrated great strength in the mayoral primaries. Jackson led the field in Atlanta with 46% of the vote; Massell, the incumbent mayor, polled only 20%; and Senator Johnson managed less than 4%. In Detroit

Young ran second to Police Commissioner John Nichols, winning 22% of the total vote. Racial polarization was evident in the voting patterns from the outset in Detroit, as Nichols won only 3% of the vote in black precincts and Young won less than 2% in white precincts (DN, 12 Sept. 1973).

The campaign in Detroit focused on the issue of crime and police behavior, but the two candidates studiously avoided exploiting the racial issue, which everyone nevertheless recognized was the defining feature of the contest. Nichols, however, did clearly stake out the lower-middle-class white neighborhoods in the northern part of the city as his home territory, and he won the fervent endorsements of the organized city employees, particularly those in the uniformed services. Nichols received only modest support from big business. Young pushed Nichols hard for supporting the STRESS unit and promised to abolish it if elected. The STRESS debate in particular emerged as the means by which the racial conflict in progress could be fought out without engaging in open racial demagoguery. If this strategem fooled no one as to the real nature of the competition, it nevertheless preserved the appearance of a contest fought on issues rather than on the basis of the color of the competitors.

Racial lines in the voting in the runoff election, however, were apparently even more sharply drawn than they had been in 1969: Young won fewer white votes than Austin had. But he took over 90% of the large black vote, enough to provide him a narrow victory margin of 14,000 votes out of a half million cast. Four blacks were elected to the nine-person council (DN, 7 Nov. 1973).

If both candidates in Detroit sought to avoid the appearance of a racial confrontation, no such inhibitions were at work in Atlanta in 1973. As Maynard Jackson's campaign developed momentum, Massell became desperate. Abandoning his reputation for racial liberalism to the winds, he repeatedly charged that Jackson's appeals to the black electorate made him a "racist." Massell told white audiences that he could think of nothing Jackson had done for the white community in his 4 years as vice mayor (AC, 10 Oct. 1973). Later in the campaign Massell adopted the slogan "Atlanta's Too Young to Die," and he began to predict "chaos" if blacks took both the mayoralty and the city council presidency (14 Oct. 1973). The mayor's bald strategy offended the press as well as many prominent figures in the white business community. "Massell's disappointing last couple of weeks in his 1973 campaign," a civic organization president commented (I-227), "put Maynard Jackson into office just like that." and the Atlanta Constitution (12 Oct. 1973), in a comment reminiscent in intent of The Boston Transcript's remonstrance of those who would exploit ethnic hatreds openly, noted: "Mayor Sam Massell acts as if he were running for mayor of a South African city which practices apartheid rather than mayor of a fully integrated American city."

In the runoff Jackson received 59% of the vote. Nine black councilmen were elected, thus comprising exactly half that legislative body. After the inauguration, Massell returned to his real estate business and withdrew from public life.

In October 1977, Jackson won a second term in office, increasing his share of the vote to 63% against a field of colorless opponents. The two closest competitors, a white businessman from the north side of town who had run far behind in 1973 and a white Fulton County commissioner, combined to take only one-third of the total vote. At least one widely known influential white man, former governor Carl Sanders, had flirted with the idea of challenging Jackson but decided not to do so. Black rule seemed firmly established in the city. On the day after the election a reporter for the *Atlanta Constitution* asked the white county commissioner if he thought "a white man could ever win another [mayoral] race." "I don't think so," Commissioner Farris replied. "I think this was possibly the last chance [10 Oct. 1977]."

The consolidation of black rule in Detroit by 1977 seemed even more thorough than in Atlanta. Young's principal challenger was Ernie Browne, a conservative black city councilman whom the press dubbed the "Black White Hope." The election marked the first time that the two major contenders for the mayoralty of a big city were both black.[5] Playing to the white ethnic and city employee constituencies, Browne attacked Young's "flamboyant style" and suggested that the mayor was nothing more than a "hoodlum street-fighter." Browne won the support of the Detroit Police Officers Association, which provided him with nearly one-quarter of his campaign funds (DN, 6 Sept. 1977). Young was backed by the black community (in the primary Browne received only 7% of the black vote) and white business interests, including Henry Ford II and the presidents of General Motors and Michigan Consolidated Gas. Key issues in the campaign were Young's affirmative action efforts in regard to the police and an increase in the property tax for the heavily black public school system. In the runoff election Young won 60% of the total vote. Ninety-two percent of the black voters supported the mayor. Browne campaigned heavily in the areas that had gone for Nichols in 1973 and won 90% of the white vote.

[5]It will be recalled that it took the Boston Irish nearly 30 years from O'Brien's first mayoralty to stage an election in which the two main candidates were both from the new ethnic majority.

SUMMARY

Postwar politics in Atlanta and Detroit have clearly been marked by the pervasive presence of racial issues. The racial cleavage has served as the preeminent factor in defining the structure of political conflict in these cities, and the problem of black power has dictated the major strategic preoccupations of both blacks and whites. Until the 1970s Atlanta politics was characterized by the continuing efforts of upper-middle-class whites to deflect both black power and lower-class white racism by coopting the one in the fight against the other. The function of the business-dominated "progressive" coalition was very much to perpetrate a politics of containment. Black political development was also held in check in Detroit politics in these years, a function mainly of the inability of the labor-led coalition to control its white ethnic rank and file. It is important to note, of course, that in both cities blacks were critical elements in biracial coalitions, but it is also important to observe that black participation was predicated largely on conditions established by whites. In return for their voting loyalty in Atlanta, blacks were virtually guaranteed by influential whites that city hall would not become a forum for a politics of unregenerate racism. Other rewards, however, were limited to the degree that they threatened or interfered with the pursuit of economic development goals. In Detroit black aspirations were blocked by virtue of the fact that blacks were normally part of the losing coalition. But even when they helped to elect Jerome Cavanagh, they were not offered a substantial share of the formal power in the form of government appointments or political office, despite the mayor's commitment to the principles of the Great Society.

With the emergence of black majorities in the two cities, several changes occurred simultaneously. Blacks were able to break from the white-dominated coalitions through which they had operated. Not only were blacks able to become essentially independent forces in their respective cities, but they lost their subordinate status in the local political system. Black aspirations no longer had to be deferred automatically if they conflicted with white goals. The emergence of black power also marked the end of the success of white containment strategies in both cities.

In a context in which racial tensions have traditionally been acute and in which much of politics was conditioned by white efforts to hold black power in check, the changes wrought by the development of a local black majority are particularly significant. And the question of how the groups that once dominated the system and orchestrated its politics have responded to this changed state of affairs lies at the heart of the matter.

4

The Psychology of Adjustment:
Perceptions of Transition
in Detroit and Atlanta

The change from white to black rule in city government is an event of extremely high salience. All during the campaign local news media are preoccupied by the implications of racial transition. On election day voters in both black and white precincts turn out in unusually large numbers (Pettigrew, 1972). During the postelection period and for several years thereafter the national media watch the new black mayor, periodically reporting on his conduct of office. Curiosity about how cities are faring under black rule attracts even foreign journalists, whose readers may scarcely have heard of places like Gary or Atlanta (see, e.g., "Atlanta," 1975). But the attention devoted to racial transition in the press only mirrors the concerns of the participants in the process, both those who have just won power and those who have been displaced. The public speeches made by black and white elites alike to civic organizations and service clubs, at political rallies and press conferences reflect high awareness of the changes that have taken place in the political order. The initial triumph of a black mayoral candidate is no ordinary event in urban politics.

The psychological adjustment of white elites in particular to the transformation of their local political world is a multifaceted process. On one level those who have been displaced focus on the process of transition itself, characterizing and dealing with its broad outlines. This involves describing the new structure of power, estimating the durability of the change in that structure, putting into words the feelings the victory of black power evokes—in general, trying to interpret the meaning and character of transition.

On other levels displaced whites focus on the person of the new mayor as a way of coming to terms with their new situation, or on the perceived impact and performance of the new regime in city hall. Although these are natural enough exercises in political observation and evaluation, both the personification of transition and the effort to establish it in the particular context of the moment may be seen as strategies to reduce and specify the magnitude of change. Thus displacement is made less ominous and practical responses are suggested, lending a tangible immediacy to what otherwise is likely to be seen as a broad, diffuse, and possibly enervating social transformation.

Underlying all three of these levels is still a fourth, namely a generalized sense of optimism or pessimism about the prospects for the city. These feelings may, of course, be related to the perception of the transition process, the mayor, and his impact, but in many cases they appear to be largely self-generated. In either event the psychological focus here is on the future.

Exploring these dimensions of psychological adjustment to transition may help in estimating the extent to which white elites will make the particular resource of their support available to the new mayor. Support in this sense refers to expressed or latent attitudes, either general or specific in their focus, which help to create an encouraging, facilitative, or permissive environment for the mayor. Supportive responses such as good will, tolerance, or patience create flexibility for a political leader, enabling him or her to make retrievable mistakes, to take initiatives with confidence, to assume risks, and to commit resources with certainty.[1]

In an environment of support there is no rush to judgment. A contrasting situation occurs, however, when prominent or visible members of the displaced group publicly assess the mayor or the political situation of the city using the language of fear, despair, or pessimism, thereby in effect withholding support from the mayor. The consequences may range from undercutting investor confidence to discouraging other whites from lending their expertise and prestige to the new regime. In an atmosphere of apprehension and distrust, the motives of the new

[1]See Gamson (1968, pp. 42–43) on such support as a resource.

mayor are constantly called into question, and his or her ability to mobilize civic energies is limited.

FOCUSING ON THE TRANSITION PROCESS

RECOGNIZING TRANSITION: COGNITIVE RESPONSES

White elites in both cities were acutely aware of racial transition, although in Atlanta the phenomenon was accompanied by a sense of loss and invested with a degree of dramatic intensity lacking in Detroit. Several prominent businessmen in the southern city described the shift in political power to the black community as "traumatic" (Viorst, 1975, p. 10; I-210). In much the same fashion as Boston Brahmins looked to the decades before the rise of the Irish, a number of Atlanta's elites idealized the Hartsfield–Allen years as a less troubling, more manageable period. In his classic study of Atlanta in the 1950s, Floyd Hunter (1953, pp. 65, 75, 113) had described the "power structure" of the city as consisting of "a relatively small group" of white businessmen with close social ties. And Ivan Allen, Jr. (1971), the city's mayor during the 1960s, has written of his generation's accession to civic leadership that "we [were] successors to the throne," likening the situation to that which he and his friends—bankers and businessmen—had faced "when we had taken over the family businesses from our fathers [pp. 29–30]." Decisions were made in the decades of the 1950s and 1960s by men who knew each other, and they were often made out of the public eye.

The mystique of this past serves as a powerfully disconcerting symbol for white elites in the contemporary city, for it throws the present shift of power into the sharpest contrast. "In the old days," one businessman recalled, "you could pick up the phone and dial the Mayor at his office or his home or his club—your club—or his friend's house—your friend's house—and you could get your business done, right there, first-name basis [NYT, 26 Feb. 1975]." The head of a major business organization (I-234) commented, "There are still some in the business community who wish Atlanta was still 1961," and a white officeholder noted in his interview (I-235) that when Maynard Jackson was first elected, "We kept hearing about 'the way it was'."

Detroit elites viewed transition in more matter-of-fact terms: "The people recognize change," said a city councilman (I-113). "We've got a city that's probably 55% black now." The period before black rule in Detroit represents no idyll before the loss of innocence. If anything, the world of the years prior to the 1967 riot was regarded as unreal and unworthy. The complacency bred by the failure to understand the depths of racial disaffection was widely interpreted, particularly within the business class, as a failure of civic responsibility. The illusions of the early part of that decade were best forgotten; Detroit elites for the most part viewed the transition and sensed a kind of progress.

That prominent whites in the political and economic spheres of Detroit did not speak of transition as "traumatic," as many of their counterparts in Atlanta did, is in some measure a product of regional differences. Notions of a racial order were simply more prevalent and more deeply entrenched in the southern city and therefore presumably all the more tender in the uprooting. Yet Detroit, like many northern cities, could scarcely claim a significantly more virtuous record in race relations than its southern counterpart. Furthermore, Atlanta's population is relatively cosmopolitan in its origins, for the city was an early destination in the corporate migration to the Sunbelt.

The differences in early reactions to transition may more importantly be a function of the differences in the structure of power in the two cities prior to black rule. Whereas businessmen in Atlanta had filled a power vacuum and reigned virtually unchallenged (Kotter and Lawrence, 1974, pp. 144–145), the Detroit political scene of that period has been described as "a balance among . . . bitterly antagonistic interests [Banfield, 1965, p. 51]." Organized labor, local business (but not the automobile corporations in those days), blacks, white ethnic groups, and city employees struggled in a politics of coalition-building to maximize their influence. The rise of black power in Detroit displaced *whites* but not a stable, cohesive *white power structure.* In Atlanta, however, the triumph of black power was seen as having displaced a ruling class, dramatically divorcing economic influence from political power. In a paper delivered before the Atlanta Rotary Club, an urban economist (Hammer, 1974) stated the matter in the following way:

> The operation of the core area economy, which is predominantly a structure owned and managed by whites, is greatly dependent upon the policies and progams of the city government, which is increasingly coming under the influence of black political leaders. That is a new situation that must be dealt with [p. 26].

An Atlanta editor (I-26) put it more starkly: "Blacks have the ballot box and whites have the money."

In light of this analysis it is not surprising that Atlanta's white elites almost universally assumed that black rule was to be a long-term prospect. Although Detroit elites were less likely to employ the economic power–political power dichotomy to characterize the nature of transition, they nevertheless harbored expectations similar to those in Atlanta of black government as a probability in the foreseeable future. For most people in elite positions in the two cities, the assumption that black rule would be a durable phenomenon was a product of a simple calculus: Government is a product of electoral outcomes, which in turn are dependent on the racial composition of the city. "Majority" and "minority," insofar as these categories relate to the city, were understood almost exclusively in racial terms. The sense that factors other than race might conceivably divide people politically in the two cities was seldom expressed. Both Maynard Jackson and Coleman Young were seen as representing majorities defined by race.[2]

The analysis of expectations concerning the durability of black rule suggests that a rationalizing response to displacement, that is, the assertion that nothing of lasting significance has changed, was relatively rare. The electoral calculus is realistic. It assumes racial bloc voting. It recognizes the importance of the mayoralty to blacks (without explicitly acknowledging the justice of a racial claim on the position), and it accepts race as the critical structural aspect of the city. Nevertheless, a few people, particularly in Atlanta, maintained during Jackson's first term that voters were more interested in getting "a good man" regardless of his race, and that "given the right situation" a white could get elected to the mayor's seat (I-116, 235, 233, 220). Two of the people who offered such analyses were would-be contenders for city hall, who might have been expected to maintain such self-serving perceptions. The more common pattern however, was to acknowledge racial transition in the respective cities as a development of significant and enduring proportions.

COMING TO TERMS WITH TRANSITION: AFFECTIVE RESPONSES

Patterns of affect in Detroit and Atlanta showed a strikingly similar progression to that which we observed in Boston: White elites in the two contemporary cities responded initially to the prospect of transition with fear, but living under black government brought gradual and widespread acceptance. Two important differences between Detroit and Atlanta elites emerged, however. In the latter city those interviewed were more likely to describe their early fears in stronger terms than did elites in Detroit. Of greater consequence, a large number of people in Detroit went beyond acceptance to express satisfaction with transition. In Atlanta such feel-

---

[2]As the *Atlanta Constitution* (7 Jan. 1974) wrote editorially during the week of Maynard Jackson's inauguration, Jackson " is the standard bearer for his race."

ings were limited almost entirely to the few whites who were actually playing roles in the mayor's network of advisers and confidants.

Atlanta respondents described their anticipation of black power as "fearful"; in Detroit people spoke of "concern," perhaps of their "apprehension." An Atlanta businessman (I-228) worried about "all-black domination the way there used to be all-white," a concern reminiscent of Rothchild's findings (1973) in his study of Kenya. A lawyer and campaign activist in the southern city (I-221) spoke of the "tremendous initial shock" whites experienced as black rule became a reality. An executive in a large real estate management firm (I-226) explained whites' "great fear": "They were afraid they wouldn't be able to call up down there any more at will." And a former high elected official (I-24) recalled his fear that a black mayor would ride roughshod over white interests.

Initial apprehension was acknowledged in Detroit, but people suggested in the same breath that it was greatly tempered. Several pointed out that ethnic succession is a historical commonplace in Detroit. Such a pattern, fashioned by Italian and Polish stock mayors and coupled with Richard Austin's unsuccessful run for the mayoralty in 1969, cushioned the shock of the victory of a black candidate. Others spoke of the quick emergence of support for the black mayor by certain prominent whites as a factor that dissipated initial concerns, and a utility company executive hedged his interpretation of white response by suggesting that although whites were apprehensive over transition, they had no problem with the particular black candidate himself (I-18, 123, 119).

If the language in which Atlanta elites described their anticipation of black rule indicates a somewhat greater emotional disturbance than people experienced in Detroit, there was nevertheless agreement that white fears in both places had diminished. Most whites at the elite level in these cities said they had accepted black government. But people in Atlanta in particular described their response in terms of "coping" with black rule, or of being "resigned" to it. In this context, then, acceptance may be seen to involve acquiescence to a condition over which an individual has little control.

Many of those interviewed explained their acceptance of black rule not as a function of its positive impacts but of the fact that it had few negative consequences. Such a pattern may be taken as a hallmark of the acceptance adjustment. Thus, a construction executive in Atlanta (I-213) noted that he accepted black rule because it has no "hindrance." And a former high elected official in Detroit (I-22), reflecting on his feelings at the time of Coleman Young's victory, recalled that having a black mayor did not "bother" him. Another figure in the northern city, the head of a large business organization (I-110), noted that the move from a white to a black mayor "hasn't hurt the city at all." And a woman who headed a civic organization in Atlanta (I-227) said: "A lot of people who thought they couldn't live with a black administration have found they can do so quite well."

Acceptance is the response of people who may believe they have little opportunity to alter their condition but at the same time do not feel seriously disadvantaged. It is a response essentially devoid of moral content, and presumes no triumph or defeat of what is "right" or "fair." Thus it represents an acknowledgement of the world as it is, a world with which one must cope with equanimity, at least insofar as one's own interests are not seriously threatened. *Satisfaction,* in contrast, is not affectively neutral; it implies a sense that a condition which exists ought to exist, that it is advantageous in some way. Although acceptance was the modal response in both cities, a substantial proportion of Detroit's elites, ranging from elected officials to corporation executives to bankers, expressed satisfaction at black rule.

Satisfaction may be expressed in terms of moral gratification or it may take more pragmatic forms. The moral version is framed mainly in the terms of democratic majoritarianism: Blacks are the majority and it is right that majorities rule. As whites as a majority were "on top" before, so it is now fair and proper for blacks to be on top. Black rule is not something simply to be coped with; it is, under the circumstances, inherently legitimate.

The pragmatic version leaves moral considerations unexpressed but nevertheless assumes them. It is the good consequences that result from rule by rightful majorities on which the pragmatic version focuses. Thus, a former Detroit politician (I-133) evaluated black rule as "a positive thing. It gave blacks in this city a sense of participation and identification with the city that no white mayor could have done." Another official argued that black rule had improved black–white cooperation, and others mentioned simply that black government was "good for the city," "the best thing that ever happened to it" (I-135, 137, 132, 138).

INTERPRETING TRANSITION: EVALUATIVE RESPONSES

In assessing the implications and character of transition, elites in the two cities tended to emphasize different aspects of the process. In Detroit the perception was widespread that working-class white ethnics in particular were bearing the major costs of transition. The white homeowner neighborhoods of northeast Detroit were seen to lack spokesmen and effective political organization. Arguing that municipal nonpartisanship foreclosed the possibility of an interracial party coalition along the lines of the Wayne County Democratic organization, elites tended to view white ethnics in the city as isolated and demoralized by black rule.

In Atlanta the question of whether particular white groups suffered disproportionately from displacement was seldom considered. To the extent that Atlanta elites thought of the white community in terms of its various parts, it was a small and dying traditional leadership element that was seen as bearing the heaviest costs. Atlanta had recently experienced a generational shift in its white leadership structure. The members of the aging plutocracy that ruled in the Hartsfield–Allen era had largely retired or died by the early 1970s. Yet a few were still active, and it was this small group, products of a dying Southern culture, who felt themselves to be especially out of place in the new order.

If elites in the two cities did not feel that the shift to black government had burdened them particularly with any specially heavy disabilities, Atlanta respondents nevertheless believed that the "rules of the game" had changed. They spoke repeatedly of the situation as a "new ballgame" necessitating adjustments. They had to learn, as one businessman noted, how to talk to "black power people." And several came to the realization that whites from the business community in particular could no longer hope to achieve positions of political leadership (I-212, 213).

It is not surprising, then, that elite evaluations of the transition process in Atlanta were extremely tentative compared to those in Detroit. The fact that the shift was peaceful—that "the lid has been kept on"—was remarked frequently (I-26, 222), an observation that suggests, perhaps, how limited white expectations were prior to transition and how simple it was to fulfill them. Others noted how difficult it was to lose power "to people you don't know" (I-217, 227), a plaint of people accustomed, surely, to Atlanta's genteel tradition of a limited and intimate ruling class. But for a banker in Detroit (I-129). "The style and color of the mayor hasn't made much difference." And a Detroit man who once sought the mayoralty himself (I-17) remarked: "People have come to understand that black rule doesn't make any difference. . . . The problems still exist. Nothing is so different. Government is government with all its limitations."

THE IMAGE OF THE MAYOR

American mayors and their cities often seem to develop a symbiotic relationship, one coming to stand for and reflect the other. Perhaps voters tend to select a mayor who is a faithful product of their particular city's culture. There could have been no mistaking the fact that Richard Daley was from and of Chicago, any more than that Richard Lugar was from and of Indianapolis. To the outside world, especially, the image of the city is in part a function of the image of the mayor. But to speak of symbiosis is also to say that mayors, as personalities and political leaders, tend to make a modest imprint on the cities they govern. Both of these attributes may worry a mayor's political opponents and provide hope for his or her supporters. When the mayor is black, the concerns of his or her white opponents and the hopes of his or her white supporters are intensified accordingly. No one is without opinions.

In certain significant ways Maynard Jackson and Coleman Young reflected the culture of their respective cities, despite the novelty of their blackness. When Jackson first won the mayoralty he was a young man, born in 1938, who came to power at the same time as those of his generation in the white community assumed business and civic leadership. His youth and urbanity fit the city's sense of itself. As Atlanta's prominent figures had always been, Jackson was well born, with parents who occupied a substantial position in the city's black community. His father was a clergyman in a major Baptist church in the city, and his mother was a professor of French at Spelman College. Jackson graduated from Morehouse College by the age of 18 and went on to earn a law degree. In 1968 at the age of 30 he began his political career, first by challenging Senator Herman Talmadge in the Georgia Democratic primary (thus becoming the first black to run for statewide office), and then by running successfully against a white opponent in 1969 for the vice mayoralty of Atlanta.

If Maynard Jackson represented Atlanta's tradition of gentility and ambition, so Coleman Young personified Detroit's labor heritage. Young's climb to the top was a struggle; his apprenticeship a long one. He was born in Alabama in 1918, but his parents moved north to Detroit when Young was 5 years old. During World War II he served in the Army Air Corps as a member of the all-black Tuskegee Airmen. Once he was arrested at an air base in Indiana for attempting to integrate an officers' club. After the war Young became a labor organizer in Detroit's auto plants and was later elected organizational secretary of the Wayne County CIO, the first black to hold such a position. In 1948 he directed Henry Wallace's presidential campaign in Michigan. This, plus his subsequent founding of the National Negro Labor Council, brought him a subpoena from the House Un-American Activites Committee. In 1960 he was elected a delegate from Detroit to Michigan's Constitutional Convention. From this experience he launched his political career in 1964 by winning a seat in the state senate. In 1968 he was chosen as Michigan's National Democratic Committeeman, and in 1973 he was elected to his first term as mayor of Detroit (Tyson, 1976).

White elite perceptions of Young and Jackson appeared to be shaped by at least four interrelated factors. These perceptions were in part a product of the mayors' personal characteristics—their career histories, styles, and personality traits. However, both men were seen not simply as mayors, but as *black* mayors. Thus race was a second factor employed widely to explain the mayors' performances and characters. A third factor determining the ways in which the two men were seen concerned certain values of the observers themselves, and a fourth factor, the perceived impacts of the mayors.

Coleman Young was seen as an experienced and shrewd politician; Maynard Jackson, despite his 4-year tenure as vice mayor, was regarded as inexperienced. "Coleman is the most qualified politician in the country," a Detroit journalist (I-119) said: "He knows the issues and he knows politics." Others who were interviewed agreed (I-116, 131). Atlanta elites, however, often commented on Jackson's lack of long experience in public life and attributed many problems to his greenness. "It would have been better to have gotten someone with a track record," an Atlanta editor (I-26) noted. And a banker (I-211) assessed Jackson as "inexperienced. . . . He's going to make mistakes, and they're very public."

Both mayors were viewed as urbane, articulate, intelligent, and even "charismatic." Jackson was likened to a preacher in the pulpit and a salesman who "could sell a blind man bifocals." His eloquence was his most widely admired characteristic. Young was regarded as "a streetwise politician," a man with a lot of "Negro political savvy."

The blackness of the two mayors was an inescapable and dominating characteristic. Indeed, Young was admired for his mastery of what were seen as peculiarly black gifts, and his blackness was viewed as having given him special advantages in his dealings with others. A department store executive (I-136) analyzed the situation in the following way: "He comes from the streets, he's smart. When he talks to labor, he talks straight, even though he's on the other side of the bargaining table." A city councilman (I-137) saw similar advantages in Young's blackness: "He has some tough union problems but he can get by with it because he's black. . . . He gets along with militant blacks."

The novelty of a black mayor was seen to induce a certain caution in those with whom he dealt. A former city official (I-133) argued that the mayor's race allowed him greater latitude in dealing with the state and national governments: "He can do and say things that cause [President] Ford and [Governor] Milliken to shake a bit. White pols are a bit scared of dealing with and shouting at black politicians. . . . When he's critical of Ford or Milliken, they are more apt to listen." Although several people criticized what they regarded as the mayor's failure to assuage the fears of the white community (I-112, 138), the more general view was that Young had managed to forge links to both the black community and white business (I-14, 110, 136). His record of militance on racial issues prior to his election was seen as having created for himself a high degree of freedom of action to deal with white business leaders. His civil rights credentials "make it possible to cooperate with the corporate community and not be seen as a stooge or toady [I-111]."

For white Atlantans, however, the race of their mayor created problems. Jackson's blackness was seen as the source of what was perceived as "touchiness" and arrogance. "Every time he gets criticized, he thinks it's racist," a former candidate for the mayoralty (I-28) noted. A city councilman (I-233) expressed the same widely held view: "A major weapon of Jackson's is to call anyone (me, for example) a racist, a polarizer, when they disagree with him or argue." Ironically, Jackson himself was viewed by some as a racist. "He thinks too much in terms of black and white," said a banker (I-219). And a county commissioner and corporate executive (I-220) believed that "everything he does is geared to helping the black people. . . . He's a racist. People are disenchanted with him."

Jackson was seen as a man caught in between two constituencies with radically opposed interests, the black community on the one hand and the white business community on the other. He was indicted for his inability to bridge the gap between these two groups as well as for his inclination to play to the interests of the former. A one-time state official, now a lawyer in a major Atlanta firm (I-229), offered an analysis shared by a large number of Atlanta elites: "I'm sure he's being pressured from all sides and he probably feels he's got to relieve the problems of poor blacks in the cities. He also wants to consolidate power in the hands of blacks who are running the city. OK, I understand that, but it's scaring the whites."

When Jackson's critics spoke of their disappointment because he had chosen to be "a black man's mayor," they were setting his mayoralty against a standard of impartiality that their own emphasis on the inevitability of Atlanta's racial dichotomy belied. The concern that whites expressed over Jackson's efforts to "do for blacks" was in part the response of a new minority uncertain of its position in the new order. "A lot of his early speeches and decisions," said a city councilman (I-224), "were attempts to make up for black disenfranchisement. He seemed to imply that whites were going to really have to pay for a long time." Moreover, such concern suggests a reluctance to accept as legitimate a black mayor's belief that he may be obliged first to address issues of special significance to those chiefly responsible for his election. The symbolic benefits a black in city hall can provide for his black constituents were scarcely acknowledged by white elites in Atlanta.

Despite the differences in the way in which elites in the two cities assessed the impact of their mayors' race, there was a notable absence in both places of racism as a mode of expression or as a context for evaluation. Pierre van den Berghe (1967, pp. 11, 23) has defined racism as the belief that physical traits are determinants of social behavior and moral or intellectual qualities. Differences in physical traits (socially defined "racial" differences) then become the basis for invidious distinctions. If Jackson's blackness was perceived as a source of his problems as a leader, it was put in such a way as to imply that Jackson personally could not handle the burden of his race gracefully. There was no analysis by white elites in Atlanta that suggested that blacks in general were not fit to govern. In Detroit elites believed that Young capitalized on his blackness to good advantage, but they did not believe that any or all blacks would or could do so because they are black. Pernicious racism, manifested by the use of racial epithets or uncomplimentary generalizations about blacks as a group, was absent from the public discourse of elites in both cities, just as it was in Boston during the Yankee decline.

There are several possible explanations for elite avoidance of racist speech and analysis. One is that the white elites interviewed were simply not racists, at least in a blatant or self-conscious sense. They either had moved beyond such modes of thinking or they had never employed them to begin with. If they were apprehensive about black rule, it was not that they believed that blacks as such could not govern competently or fairly but rather that power had shifted to a once politically deprived group lacking both experience in government and a record of political leadership. Therefore apprehension arose from the unpredictability of the situation.

A second explanation is that racism was regarded as imprudent. Power breeds respect, or at least the appearance of deference. The resort to racism constitutes a potentially dangerous provocation for a new minority to consider.

A third explanation suggests that racism was displaced by the use of neutral-sounding criteria of evaluation that function in effect as code words, masking unacceptable modes of thought and expression for both users and their audience. In Boston, we may recall, the conflict between Yankee and Irish was conceived in public discourse not so much as an ethnic struggle but as one between the forces of reform and machine. Explicitly anti-Irish and anti-Catholic pronouncements were shunned, and indeed, disavowed by a substantially large proportion of the Brahmin community. This pattern seems particularly apparent in Atlanta. Whereas the Boston Irish were attacked for their commitment to a politics inimical to the principles of good government, Maynard Jackson was criticized for his lack of managerial experience and administrative ineptitude, as the following quotes show.[3]

> I rate him low on management. I attach a lot of importance to that. He has no experience in business [former high elected official, I-24].

> He's a good salesman [but] not an effective administrator [banker, I-211].

> Jackson can't make decisions in the sense in which an executive should make decisions [city councilman, I-235].

> Lawyers are bad administrators. One advantage of businessmen in politics is that they have administrative experience. That's a problem with Maynard [department store executive, I-212].

Such business attributes may be important for a political executive, but the standard of managerial experience is one shaped by Atlanta's past, when mayors were white businessmen, rather than by the different conditions of the present. Naturally enough, perhaps, white elites projected their own values onto a conception of the mayoralty, but by elevating managerial experience to a position of primacy, they established a narrow notion of the range of functions of the elected political executive. That such a role might include political leadership—mobilizing, innovating, symbolizing, mediating—was scarcely admitted. (In contrast, Coleman Young was generally praised for his leadership qualities and his forcefulness [I-14, 132, 131, 118, 119, 121, 114].) Hence, it is not simply because Jackson sought to play a *black* leadership role, thereby hovering over some implicit boundary between the acceptable and the unacceptable, that the use of the standard of administrative experience appeared to take on the coloring of a masking strategy. The analysis of the neutral-sounding standard as a code word is lent credence by the fact that few blacks could possibly meet it, for blacks have generally been denied major managerial roles in government and business.

None of these three explanations for the absence of racism in the evaluations of the black mayors—that those interviewed were not racists, that they considered racist analyses too risky in their position, or that they sublimated or masked their racism—is so compelling as to elimi-

---

[3]These are apparently characteristic indictments in black-mayor cities. White elites in Newark accused Kenneth Gibson, the city's black mayor, of being a poor administrator (*NYT*, 10 July 1977), and the same was said of Gary's Mayor Richard Hatcher (see Poinsett, 1970, p. 133).

nate the possibility of the others. The pattern of discourse was probably a function of all three. What is important, of course, was that racist analysis and language were not considered acceptable in either city for whatever reason. Despite the salience of the racial factor in the assessments of the black mayors, the absence of racism must be seen as an important, if possibly subconscious, contribution by white elites to the maintenance of comparative racial amity and to the moderation of what racial conflict does occur.

By eschewing racist analysis, white elites critical of the black mayor diminished the severity of their minority situation. Although they generally assumed that black rule would be a durable phenomenon, focusing their discontent on what they saw as the shortcomings of the individual incumbent rather than on the perceived defects of the race as a whole opened the way to accepting both the principle of black rule as well as future black mayors. A focus on the individual mayor also helped to create more manageable strategic options for the white minority than an analysis that framed discontent purely in terms of a white struggle with a black majority of questionable competence and intentions. If the electoral calculus made the search for a white candidate a futile enterprise, at least the white minority could hope to find a more acceptable black in the future.

THE PERCEIVED IMPACTS OF THE BLACK MAYORS

The first actions taken by political executives after they assume office are likely to be heavily fraught with symbolic significance. These early initiatives establish a set of expectations about the new officeholder's style and predilections. Subsequent activity is compared to that taken in the "first 100 days" or its mayoral equivalent. It is through these intial actions, too, that a politician begins to pay off or reassure the important elements in the constituency that elected him or her. For all these reasons, a newly elected mayor may be expected to take his or her first steps with greatly calculated care. For both Maynard Jackson and Coleman Young the earliest and most visible actions involved efforts to challenge and control the police, to increase opportunities in government for blacks, and to foster the economic development of their respective cities. It was to these areas that elites generally referred in offering an evaluation of their mayors' impact on the city.

One of the common questions asked of the elites who were interviewed for this study was to assess the general performance of their mayor. Follow-up questions sought to explore the specific criteria on which the general assessment was based. Atlanta elites were divided in their general assessments, although those who expressed "disappointment" in Maynard Jackson were a substantial majority. Of those people willing to make a general assessment, those who offered positive evaluations came largely from the political sphere: a former mayor, several city councilmen, and professional campaign managers (one of whom was close to the mayor). Negative assessments were concentrated among business elites. In Detroit Coleman Young was almost universally praised in general terms by people from all sectors of the elite group. Table 4.1 presents a breakdown of the specific issue areas and the manner in which mayoral impacts were assessed.

The dilemmas involved in policing a city probably create more intensely felt racial animosity and involve more people in both racial communities than any other of a host of perennial urban problems. Unlike other sensitive issues, such as busing, which primarily galvanizes the white community, or the desegregation of residential areas, which mainly affects specific neighborhoods, the twin issues of the conduct of the police and the control of crime are matters of intense and general concern. Starkly put, whites worry about the threat of black crime and the possibility that police efforts to control it will be hindered. Blacks, worried themselves about crime, fear the repressive capacities of overwhelmingly white police forces and the dangerous implications of prevailing racial stereotypes regarding the urban criminal. In this context a challenge to the police offers a new black mayor a tempting and dramatic opportunity to demonstrate to the black community that having a black in city hall makes a difference where police behavior is concerned. At the same time, however, such a challenge evokes great anxiety among whites.

In both Atlanta and Detroit incidents involving the mayors and their police forces emerged as the most salient events defining the character of the mayors' performances. Both mayors had campaigned heavily on the shortcomings of the police services, and once in office each sought to assert control over the police. Young was the more adroit in his challenge. In his inaugural speech he sought to establish from the outset an uncompromising position on crime—thereby allaying white fears—by ordering "pushers, ripoff artists, and muggers" to "hit the road [NYT, 6 Jan. 1974]." Against this background he acted less than 2 months later to abolish the highly controversial tactical police unit called STRESS (Stop The Robberies, Enjoy Safe Streets), which in its 3 years of existence had been responsible for the killing of 17 Detroiters, most of them black. Besides fulfilling his campaign promise to do away with STRESS, a unit to which the police and various organizations of elderly

TABLE 4.1
Perceived Mayoral Impact in Issue Areas, Ranked by Frequency of Mention

Positive		Neutral		Negative	
Atlanta (1975)					
1. Administration	8	1. Economic development	6	1. Police	19
2. Housing	4	2. White flight	2	2. Economic development	11
3. Economic development	3	3. Administration	1	3. Administration	9
4. Neighborhood preservation	2			4. Race relations	6
4. Crime	2			5. Garbage	4
5. Other	6			6. Airport expansion	2
				6. Tax increase	2
				7. Other	2
Total mentions	25		9		55
Total respondents answering assessment questions:			33		

Detroit (1976)					
1. Administration	14	1. Economic development	2	1. Administration	9
2. Economic development	10	2. White flight	1	2. Police	6
3. Police	7			3. Race relations	2
4. Race relations	5			4. Other	3
5. Other	1				
Total mentions	37		3		20
Total respondents answering assessment questions:			31		

86

whites were deeply committed, Young also pledged to increase minority representation on the force, 85% of which was white. In the spring of 1975, however, the retention of new black policemen, whose presence on the force had risen from 15% to 27%, was jeopardized when declining tax revenues forced the city to plan layoffs among its uniformed services (Tyson, 1976). Since layoffs are normally determined by seniority, the newly hired blacks would ordinarily have been the first to go. To prevent this Young ordered the exemption of that group of black policemen from the layoff order, subjecting the remainder of the force to the seniority rule. Resentment among the police and in many segments of the white community was intense (I-116, 120, 131, 137).

Young began to gain credit in the white community, however, by his actions during and after a near riot in the city in July 1975. When angry young blacks gathered two nights in a row to protest the shooting of an 18-year-old youth by a white bar owner, Young himself went to the scene to talk to the crowd. After the trouble subsided, he repeatedly praised the police for their professionalism and restraint. A labor leader (I-14) commented on the incident: "He was out there on the front line. This changed a lot of policemen. He's gotten their cooperation. He's gained a growing confidence among the white community because of that incident last summer." The head of a large business organization (I-110) assessed the impact of the mayor's response in the same way: "Respect for Young shot up with the incident out on Livernois. He went right out there and walked the streets and rattled some heads."

Young managed both to signal the black community by his actions in regard to the police and to maintain or regain at least minimal credit in the white community. The mayor's relationship with the police was seen as the single most obvious indicator of the quality of his performance. Despite conflicting pulls of the two racial communities on the police and crime issues, it was felt on the whole that Coleman Young had done a reasonably good job. Even a high official in the Detroit Police Officers Association (I-116), a critic of the mayor, conceded: "The relationship between Young and the police is about as good as it could be, given the fact that he came into office knocking the police." What is important about the pattern of sentiments regarding this particularly delicate issue area is that it suggests a relatively permissive and flexible white elite attitudinal matrix within which the mayor may act. The mayor in Detroit was able to operate in a situation in which he could regain credit and approval, even if he had taken actions perceived by many as threatening or injudicious.

In Atlanta, Jackson's dealings with the police became an even more important point of departure for white elites in assessing the mayor's performance than the analogous events in Detroit were. Both his ac-knowledged critics as well as those who claimed to be friendly to the mayor agreed that he had mishandled his efforts to gain control of the police. The consequence of Jackson's strategy was to overdraw what modest credit he had at the outset in the white elite community.

Jackson had campaigned against Police Chief Inman in part in re-sponse to black claims that Inman ran a "racist" force (NYT, 24 June 1974). After his election Jackson sought to fire Inman. The police chief refused to go, claiming in court that he could not be removed during his 8-year term. While the issue was in the courts, Jackson, under authority granted in the new charter, established the post of Commissioner of Public Safety to oversee both police and fire operations. The county court then upheld Inman's claim to a job on the force but at the same time added that the mayor could not be prevented from exercising his power of direction and supervision over Inman (AC, 16 May 1974). Jackson moved promptly to assign Inman to head the division of police services, a position that effectively removed the former chief from the inner circle on the police force.

The court also upheld Jackson's authority to establish and appoint a Commissioner of Public Safety. Several months thereafter Jackson ap-pointed Reginald Eaves, one of his aides, to the "superchief" position. Eaves, an old college classmate of the mayor's, had earlier returned to Atlanta from Boston where he had worked in the corrections system. Members of the white business community charged "cronyism" and protested the appointment of a man without police experience (AC, 3 Aug. 1974).[4] Eaves's appointment was nevertheless approved by the city council, 12–6, as three white and all nine black councilmen supported the mayor's choice.

In the succeeding 6 months various "scandals" involving Eaves came to light: an incident of nepotism, the alleged misuse of police property by one of Eaves's men, the hiring of an exconvict as Eaves's personal secretary, and so on (AC, 16 June 1975). Downtown business leaders renewed their protest against Eaves, demanding his resignation. One informant close to the mayor (I-231) suggested that privately Jackson was ready to let Eaves go, and Eaves himself apparently believed that he should leave. However, when Wyche Fowler, the white president of city council and a man with mayoral ambitions, took up the demand for Eaves's resignation, the mayor was moved to defend his commissioner. Eaves remained in office.

[4]It was during this controversy that Jackson suggested that some of the opposition to Eaves was racist. Jackson claimed that this was the only time he had called anyone a racist ("Can Atlanta Succeed?", 1975, p. 112).

The mayor's appointment and subsequent defense of Reggie Eaves was not regarded by Atlanta white elites as an isolated mistake of narrow dimensions. "That's one decision that's had bad long-range effects," noted a former candidate for mayor and member of the business community (I-28). "Eaves killed Maynard politically," said another businessman, the head of a major construction and development company (I-213). Others suggested that the appointment polarized the city racially, that it hurt the city's chances to attract business, and that it destroyed Atlanta's "image" (I-22, 27, 29, 231). That a single decision could become so important as a negative measure of Jackson's performance indicates that his credit—a function of the scope and intensity of the matrix of supportive attitudes—was relatively low among white elites from the outset and easily withdrawn. The case suggests a situation in which critics have prejudged the mayor and have simply waited for an excuse to pounce. Having found it, they could give vocal free rein to their accumulated misgivings.

If challenging the police served as the most visible reassurance the black mayors could make to their black constituents, economic development efforts may be thought to perform a similar function vis-à-vis the white business community. The two mayors worked particularly hard in this area, but the fruits they harvested in terms of positive white elite evaluations differed in their respective cities. Both mayors took office in a period of national economic recession, but each city felt the impact and suffered in its own way. For Atlanta the economic downturn meant the end of an extraordinary growth period, especially in real estate and building construction. By the mid-1970s Atlanta entrepreneurs discovered that they had overbuilt. The supply of office space, hotel rooms, and convention facilities far outstripped demand. The number of jobs in construction declined from a peak of 50,000 in the early 1970s to 31,500 in 1976 (NYT, 9 Jan. 1977). For a city that had cherished an image of itself as a boomtown, economic stagnation was a shock to morale. As a consequence, Atlanta's white elites, particularly in the banking and business communities, were preoccupied with economic recovery during the first half of Maynard Jackson's first term.

Although concern about the local economy was also deeply felt in Detroit, the roots of the northern city's difficulties were quite different. Whereas real estate speculation and the rising cost of money had brought Atlanta's economy to a halt, a decline in purchasing power among the nation's consumers had badly affected Detroit's auto industry. In 1975 unemployment in the city rose to 15.6%, the highest among the nation's 24 largest metropolitan centers. Between 1969 and 1973 the number of people employed in Detroit declined by 20%. Since more than one-third of the city's tax revenues is derived from city income and state sales taxes, Detroit's fiscal condition hovered near the edge of bankruptcy *(Report of the Mayor's Task Force,* 1976, pp. 1–3).

The economic situation of the two cities impelled white elites in the business and banking sectors particularly to place a high priority on local government efforts to aid in the stimulation of job development. Jackson and Young were both active in such attempts. Led by Chamber of Commerce officers, teams of businessmen accompanied by their respective mayors traveled repeatedly to other cities in search of industries interested in moving. Both mayors were thought to be skilled in salesmanship, but only Young was felt personally to be an asset. "When we want assistance in selling," a Detroit Chamber officer (I-110) noted, "we go to Coleman. When we bring VIPs in from Washington or New York or the trade associations, he's the host. Coleman believes in this city. No one has a greater commitment to this place." According to another Chamber spokesman (I-11), Young "makes a good impression for our business image. He sells the city."

Although Jackson's participation in such efforts was acknowledged by Atlanta elites and his personal skills recognized, his mayoralty was nevertheless regarded as a liability in the business world. A newspaper editor (I-29) offered a common view: "We were beginning to get national business headquarters here, but now interest is dead. You can't call people racists every day and have industrialists move to town. He didn't understand this." In addition, several people argued that the appointment and retention of Reggie Eaves had frightened off potential investors (I-21, 219). Others criticized his efforts to increase black job opportunities for its dampening effect on development: "The blacks want jobs. Someone has to do projects to make jobs. But here these people are damaging the very things that would make more jobs. The blacks say you can't build unless you hire so many blacks. So these things can't be built [I-215]."

Neither Jackson nor black government in general were blamed for the city's economic stagnation itself. About half those interviewed suggested that the city was caught in the throes of a national slump over which Jackson could have little influence. For these elites Jackson was viewed as a victim of circumstances. Those elites who argued that Jackson's presence was harmful did so not in the belief that he was responsible for the onset or depth of the recession in Atlanta, but rather that he represented a significant barrier to swift recovery.

Both mayors were also active in efforts to foster development and expansion of local industry and capital. Jackson was cited several times in interviews for his facilitation and promotion of a downtown housing project financed by a large local consortium, and he was regarded as friendly toward airport expansion. But his efforts in such areas were seen as neither visible nor persistent enough to earn him a reputation of sensitivity to business concerns.

A mayor's visible actions may be taken as signals to particular constituencies. Although perceptions of his or her sensitivity to the interests of a given constituency are in large part a function of the mayor's activity, such perceptions are also to some degree a product of the willingness of the audience to pick up those signals and interpret them in the manner in which the signaler wishes.

For a black mayor to participate in out-of-town trade missions with the Chamber of Commerce crowd or to appear at various business forums with bankers and executives, as both Young and Jackson often did, is to court significant political risks. By pursuing such activities a black mayor walks a thin line between what may be seen in the black community as "legitimate" relations with business and "selling out."[5] Thus to take a black mayor's overtures to the business community purely at face value and judge them insufficiently energetic or too infrequent is to deny the relevance of the associated symbolic and political costs he or she risks. Such a response cannot be regarded as particularly sympathetic. It is not unreasonable to argue that if Atlanta elites recognized the potential costs the mayor might incur with his black constituency, then whatever public efforts Jackson made on behalf of business interests should have won praise, regardless of the actual amount of activity. But white elites there did not suggest that the mayor did what he could (they therefore did not seem to understand his dilemma). Their acknowledgment of his activity was at best neutral in tone.

In Detroit, however, Coleman Young's attempts to foster local business interests were regarded as responsive and helpful. A Chrysler Corporation executive (I-118) commented:

> He works diligently with the business community to keep us where we are. He worked very hard to get us to keep our Jefferson Avenue plant, and so far it looks like we will. It's an old, uneconomical, multistory plant. . . . Coleman worked for a state law that allows a plant to renovate but keep its tax assessment the same for ten years. This will make a company think twice about building a new place. We took advantage of the new law with our Mack Avenue plant. Staying meant thousands of jobs here.[6]

[5]A prominent black businessman in Atlanta (I-218) who had been helpful to the mayor advised Jackson early in his tenure: "You don't have to marry the Chamber crowd; just shack up with them every now and then."

[6]The law is the State of Michigan's Plant Rehabilitation and Industrial Development Districts Law of 1974.

Executives in General Motors and Michigan Bell Telephone each gave similar accounts of Young's attempts to help their corporations. And an officer of Detroit Renaissance (I-124), an organization devoted to business promotion, noted:

> Coleman Young sits on the board of Detroit Renaissance. His priorities are Detroit Renaissance's priorities. We agree. . . . The investment [for Renaissance Center] was made at a point when it was assumed there was going to be black leadership. There was no panic in the business community.

It is, of course, possible that Young was more successful than Jackson in sending positive signals to the white business community. But even if this were the case, that community was also more willing than its counterpart in Atlanta to interpret those signals favorably.

A third area that elicited a large number of evaluations involved specific administrative actions taken by the two mayors. The majority of these assessments focused on the quality of mayoral appointments. Opinion in both cities was divided. Patterns of praise and criticism, with only a few exceptions, seemed without overt or latent racial content. One of these exceptions involved the approval expressed by several people in each city for their mayor's efforts to involve more blacks and women in responsible positions in city government. But most assessments of mayoral appointments focused on the perceived skills and characteristics of the appointees. Jackson, for example, was criticized for "bringing in outsiders," a reference primarily to his appointment of Jules Sugarman, once in the Lindsay administration in New York City, to the post of city administrator. Others, however, praised that same appointment, seeing in it a sign that Atlanta could still attract competent white people to public service. And because the man had a good reputation as an administrator, his appointment was also taken by some as an indication that Jackson had recognized his own administrative shortcomings.

If, on balance, elites in both cities were more critical of mayoral appointments than not, individuals nevertheless had their various favorites, both black and white, in city hall. These favorites stood in contrast to the "cronies," the "politically oriented types," and the incompetent and unqualified who were thought to make up the rest of the appointees brought into the administration. By all indications, strategies of criticism of mayoral appointments seemed to reflect the perceptions of well-informed observers of city politics rather than the plaints of a group displaced. In this sense the commentary differed little from what one might expect to find in any city under any mayor.

Administrative performance of the mayors with regard to other issues besides staffing the major departments and boards elicited assessments determined largely by the broad economic interests of the observers. Banking and corporate officials in Detroit, concerned about the city's fiscal solvency, were particularly impressed by what they saw as the mayor's courage and decisiveness in cutting expenditures and ordering layoffs to offset the municipal budget deficit.[7] Several labor leaders praised Young's fair manner in his dealings with city employees. In Atlanta real estate developers and construction executives sharply criticized Maynard Jackson's orders to award some city contracts to black-owned firms, arguing that these efforts were delaying progress on airport expansion and the construction of the city's mass transit system. When Jackson refused to support a large wage increase for the city's predominantly black sanitation workers in 1977, however, precipitating a prolonged strike, white elites issued statements of support for the mayor's stance in dealing with the union (*NYT*, 5 April 1977).

In general neither mayor was perceived as a particularly skillful administrator. But we may observe once again that what distinguishes the assessments of the impact of the two mayors in their respective cities is that Detroit's elites were willing to accept, indeed occasionally praise, Young for other qualities. By doing so they entertained a broader vision of the nature of the mayoral role, thereby relieving the mayor from the constraints of rigid expectation. Thus it is possible to conclude that in his first 2 years at least, Mayor Young enjoyed a latitude for action that was a result of the context of basically positive elite support. Jackson, however, like the Irish mayors judged against the reform ideal, was expected to conform to a more narrowly defined standard, which he did not meet.

THE GENERAL MOOD: THE NECESSITY OF OPTIMISM

No matter how they feel privately, city dwellers are likely to react defensively to any public suggestion that their town has sunk irretrievably into the quagmire of urban problems. White elites in Detroit and Atlanta are no exception to the rule. "Detroit ain't down yet!" argued Max Fisher, a major industrialist and financier, on the opinion page of *The New York Times* (3 Oct. 1976). And in a short position paper, Thomas Hamall (n.d.), executive vice president of the Atlanta Chamber of Commerce, wrote: "Atlanta is not a city in crisis; it's a city facing a crisis and responding to it in many positive ways. Look up, Atlanta!" Such profes-

[7]By late 1975 Young had cut the city payroll by reducing the ranks of municipal employees from 25,000 to 21,000 (*DN*, 26 Oct. 1975).

sions of faith and exhortations are, indeed, common in the two cities; moreover, they seem largely unrelated to people's views of transition, the mayor, or his performance. They seem in some sense to be a condition of survival.

Among those with whom the future of the city was discussed, majorities in both cities—16 out of 20 in Atlanta, and 14 out of 19 in Detroit—expressed profoundly optimistic views. The largest number of the optimistic assessments can only be said to be generated by booster impulses or blind faith. Professions of confidence emerged as if by compulsion, unsupported by reason or analysis.

Things are going to work out [former elected official in Atlanta, I-214].

I have a firm belief that Atlanta's going to come out of this [department store executive, I-228].

I don't know how we're going to get on top of our problems, but we'll do it [Atlanta county commissioner, I-230].

I think Atlanta is moving forward to fulfill its great destiny. I fully believe that [city councilman, I-223].

I'm optimistic about Detroit and continue to be. . . . We've got a great city [former elected official, I-122].

I'm optimistic. I'm a dedicated Detroiter [corporation executive, I-125].

If there's a solution to the problem of urban life, we'll find it here in the city of Detroit [corporation executive, I-132].

For others, optimism is based on economic projections. Massive building projects in the downtowns of both cities, the construction of a rapid transit system in Atlanta, and employment projections are among the factors that provide hope for a healthy future. Still others feel that the nature of their investments in the city permits no other outlook: "We have to be optimistic," a utility company executive in Detroit (I-123) said. "We have a three and half billion dollar investment in the state of Michigan." And a city councilman in Atlanta (I-235), a full-time corporation lawyer, pointed out of his office window in downtown Atlanta and noted: "There are enough people within walking distance of us who have a big enough investment in Atlanta to make sure the city will make it."

The optimism of elites in Detroit and Atlanta is the reflection of a pervasive national trait: Americans are characteristically an optimistic people (Andrews and Witney, 1976, p. 326; Watts and Free, 1976). Yet the hopeful projections regarding the future of the two cities are in no sense rationalizations; although such optimism may be the product of blind faith, it seldom assumes a false current reality. Rather, the optimism emerges from an explicit recognition of the problems of urban life and racial transition and represents a confidence in the collective ability to overcome whatever obstacles impede the fulfillment of the destinies that Americans have always expected.

One may understand the optimism of these elites as a kind of necessity. In part it stems from the need to protect self-esteem: Who, after all, wishes to be associated with a losing proposition?[8] And how can one justify one's life if the future is not meaningful? In a related way optimism is essential to the protection and fostering of one's economic interests. To the extent that one believes the health of the city to be dependent on continued investment by private capitalists (thus creating jobs and tax revenues), any public gloominess regarding the future will be taken as a warning by those with money to spend or lend. Optimism thus becomes essential for maintaining the image of the city in the business world.

The optimism we observe among elites in Atlanta and Detroit may be seen as a modest resource for the cities' mayors, for in a sense such beliefs represent a commitment that may possibly be called into play. A belief that "the city is going to make it" or that "things will work out somehow" is both prelude and incentive to efforts to achieve the goals of survival and success. Furthermore, the doubts of outsiders regarding the economic and political viability of the city have the effect of unifying people in a defensive response. To refuse under these circumstances to join efforts to ensure the survival of the city, even if it is a black mayor who organizes those efforts, is to belie one's optimism.

ELITE ATTITUDES AND THE ROLE OF THE PRESS

Although editorial policies and reporting strategies of metropolitan dailies may not always accurately reflect nor greatly influence local elite opinion, newspapers nevertheless represent the most visible and widely distributed medium for the promulgation of certain elite views of a mayor and his administration. For other elites, newspaper opinion provides a foil against which to react. Furthermore, to the extent that the local press has any nationwide visibility—particularly for potential investors and other members of the press—it helps to form a particular view of the city and the mayor. The stance the press adopts thus serves as an impor-

[8]Two of those in the small group of pessimists were men whose political careers had recently ended in defeat. A third was preparing to leave the city for greener professional pastures and did so shortly after the field portion of this study was completed.

tant element in shaping the general attitudinal context within which a mayor must act.

In Detroit the two major daily newspapers adopted a generally supportive stance in both their editorial policies and their reporting strategies, a fact that provided Mayor Young with substantial breathing room. A content analysis of editorials that mentioned the mayor and his administration, the results of which are reported in Table 4.2, shows that the more liberal *Free Press* (which endorsed Young in 1973) as well as the more conservative *News* (which endorsed Nichols) were seldom critical of the mayor. Actual praise began to fall off slightly as the mayor's first term proceeded, but it still remained substantial. Inspection of month-by-month trends in editorials shows that editorial opinions were generally a function of specific issues rather than of any preconceived notion of the character or ability of the mayor. Thus, patterns of criticism and praise fluctuated with the newspapers' judgments of the mayor's particular behavior and positions. Neither Detroit paper carried out any reporting strategy designed to cast doubt upon Young's abilities or to enhance his strength (I-12, 13, 119).

A former elected official (I-133) explained the gentleness of the press by arguing that the papers were "faint-hearted about taking on black political power. They say 'What the hell' and aim their circulation at the suburbs. . . . The papers don't even criticize Coleman when they ought to." Journalists, however, suggested that the mayor had done nothing to elicit sharp criticism. His administration was relatively free of scandal and serious errors of judgment. Thus, to the limited extent that the Detroit newspapers contributed to the matrix of attitudes within which the mayor operated, it was for Coleman Young a supportive and permissive contribution.

The two daily Atlanta papers showed a different pattern. Although the editorial positions of the more conservative *Journal* were on balance slightly more critical than supportive, its reporting strategy revealed no particular biases. The *Constitution*, however, while generally positive in its assessments of Jackson's administration on its editorial pages, simultaneously carried out a highly visible and profoundly critical front-page reporting strategy. In the spring of 1975 the paper ran a seven-part series entitled "A City in Crisis," in which the mayor's shortcomings featured prominently. Each installment began with the following Editor's Note:

> Throughout the Sixties, Atlanta was Camelot. Spared serious racial turmoil and blessed with experienced leadership, the city became a great center of commerce and a mecca for emerging blacks. Today, political power has shifted. New leadership wrestles new problems. There are tensions among the people. Camelot has faded. What's happening to Atlanta? Will the dream survive [23 March 1975]?

TABLE 4.2
A Content Analysis of Newspaper Editorials in Atlanta and Detroit, 1974 and 1976

	Atlanta						Detroit					
	Constitution			Journal			News			Free Press		
	Positive	Neutral	Negative	Positive	Neutral	Negative	Positive	Neutral	Negative	Positive	Neutral	Negative
1974	47%	15%	36%	24%	46%	30%	53%	20%	25%	47%	38%	16%
N	30	10	23	17	32	21	21	8	10	30	24	10
1976	38%	51%	9%	34%	31%	34%	41%	25%	33%	38%	57%	6%
N	21	28	5	12	11	12	20	12	16	26	39	4
All	43%	32%	23%	28%	41%	31%	47%	22%	30%	42%	47%	11%
N	51	38	28	29	43	33	41	20	26	56	63	14

97

The reporting team went on in its seven stories to cite such problems as white flight, the lack of experienced and "cohesive" leadership, racial polarization, crime, and scandal (mainly Eaves's improprieties) as reasons for a pessimistic prognosis. Much of the blame for the city's condition was laid at the feet of the mayor. "In his 15 months in office," the paper wrote (23 March 1975), "Maynard Jackson has come to symbolize to many white businessmen the troubles in Atlanta."

Officials in the Chamber of Commerce and Central Atlanta Progress (an organization of downtown business interests), as well as department store executives and other businessmen, viewed the *Constitution* series as misleading and dangerously overstated (I-216, 234, 228, 212, 232, 222). In a speech to a local business club, a one-time president of the Jaycees (Bryant, 1975) summarized common fears:

> [A]s a businessman, deliver me from the thrust of the reporting exemplified in those seven articles about the city. The word goes down: Atlantans are questioning and lack confidence in their own city. Reporters and editors around the world relay the word that Atlanta is up for grabs. And, bango! all the Forward Atlanta money, foreign and domestic trade mission efforts, industrial development efforts and literally millions of dollars invested in national and international public relations and advertising are all washed down the drain by the Sunday morning headlines.

Many business leaders professed to be unable to explain the newspaper's "vendetta" against Jackson, which, more than its editorial pages, was taken to be the measure of its true opinion. Comments by several prominent figures suggest that plain fear over the loss of white power was a prime motive force behind the series. A long-time political insider on the Atlanta scene (I-217) observed: "The newspapers are such sorry things now. Tarver [the publisher of the *Constitution*] isn't worth talking to. They used to be able to do anything they wanted, but now they can't." And a former department store executive (I-212) commented of Tarver: "He acts like somebody took his candy."

The split between the newspaper, traditionally counted among Atlanta's most influential institutions, and the business community over the issue of the paper's reporting stance toward the mayor and the city suggests that criticism of the mayor that also implicates the city oversteps certain implicit boundaries. Business spokesmen in Atlanta attacked the *Constitution* not so much for imposing the constraint of critical scrutiny on the mayor but rather mainly for its possible effects on the city's economic prospects, given the form the newspaper's strategy took.

To a large extent the distress of the business community over the "City in Crisis" series muted the impact within the city of the *Constitution's* assault on the mayor. If the matrix of elite attitudes within which the mayor operated was not more or less universally supportive, as it was in Detroit, then at least the mayor's major antagonist was isolated from its own natural allies. In addition, the daily editorial fare Atlantans received from the *Constitution* was on the whole at odds with its more concentrated reporting campaign. In the end the paper rather than the mayor lost the most credit.

THE ALLOCATION OF WHITE RESOURCES: SUPPORT AS A RESOURCE

If we limit our notion of support to mean the existence of generalized good will toward the mayor, then we may conclude that such support existed only in Detroit. Given that city's history of racial turmoil, this is perhaps remarkable enough. But there are, in addition, other supportive factors present in both cities that contributed to the ability of the two mayors to govern. In both places white elites essentially accepted the fact of racial transition. That such acceptance was underlain in part by a realistic appraisal of electoral possibilities (given the demographic balance) suggests a practical commitment to majoritarianism, a basic democratic form. The commitment to majority rule is no mean safeguard of the winner's right to govern. The perceived capacities of the individuals involved aside, the black mayors' claims to city hall were accorded therefore a certain legitimacy. Another supportive factor was the absence of overt racism or racist analysis, a factor that must have served to limit the depth of racial animosity, and yet a third supportive element was the widespread optimism of white elites regarding the future of their cities.

Atlantans were to be sure substantially less generous in their support of their mayor and all that he represented than were Detroiters. Atlantans maintained a narrower conception of "mayorship," were more nostalgic for the past, less patient in their judgment of the mayor's performance, and less prone to see the mayor's blackness as in any way advantageous to him or to the community. Nevertheless, Maynard Jackson did not face an unrelenting monolithic pattern of noncooperation and rejection. He was the object neither of fear nor of revulsion. Atlanta's white elites resignedly accepted his presence. Although the matrix of attitudes within which he operated cannot be characterized as permissive or facilitative, neither was it so confining as to preclude governance. In contrast, in Detroit white elite attitudes formed a comparatively flexible context for mayoral activity, and Coleman Young maintained a stock of credit among whites in significant positions that provided a high level of security.

5

Strategic Adjustments to the New Order

At Maynard Jackson's first inauguration, an undeviating affirmation of conventional symbols of respectability, the Atlanta Symphony Orchestra played the choral movement of Beethoven's Ninth Symphony, and the new mayor's soprano aunt entertained the crowd with a collection of arias and classical songs. In his inaugural address Jackson spoke not of policy and politics but of love. More than 90% of those in attendance at the Civic Center ceremony were black, most of them middle-aged (AC, 8 Jan. 1974). Although both the outgoing mayor, Sam Massell, and then governor Jimmy Carter came to the event, few members of the acknowledged white "establishment"—principally downtown business figures—were there. Indeed, in the succeeding weeks, members of that group were publicly relatively silent about the mayor and their feelings toward him. Much of the attention of the white community focused tensely on the city council, for its members, immediately upon taking office, engaged in a minor racial struggle over the proper scope of the powers of the white council president.

The celebration that accompanied Coleman Young's accession to city hall was more flamboyant and less self-consciously cautious. The official party lasted 3 days. A popular Motown entertainer sang for

Coleman Young and his guests at Cobo Hall, and the new mayor spoke toughly in street argot about the problem of crime. On the day after Young took his oath of office, business and labor leaders gathered for an inaugural luncheon at which the main speakers were Leonard Woodcock, then head of the UAW, and Henry Ford II. Governor Milliken and 3000 others, mostly white, were in attendance. Their expressed purpose was to pledge their support to the new mayor.

The reserve of Atlanta's white elites during Jackson's inauguration week was symptomatic of the uncertainty they felt about their role as a new minority. For the most part strategic responses to transition in Atlanta reflected their doubts; there was a certain formlessness and lack of clear purpose in the way in which white elites sought to deal with the mayor and the transition he represented. After the initial failure to be prominently in evidence at the inauguration, much of the effort to establish a place in the new order involved simply the search for ways of opening and maintaining communication with the mayor.

Although Detroit's white elites were, after the gesture of the luncheon, no more organized in their strategic responses to Coleman Young, they approached the transition with a firmer sense of their own civic and political responsibility. But in part they acted as they did for having been schooled in the great riot of 1967, an experience their counterparts in Atlanta never had.

The strategic responses of white elites in the two cities were earlier defined as those plans and actions designed to promote and protect one's interests under changed political circumstances. Empirically, these responses mainly took the form of trying to establish or maintain access to city hall, finding various political roles for oneself in the new order, planning for the political future of the city, arranging the disposition of institutional and economic resources, and seeking various oblique ways by which to change the new balance of power. Analytically, these diverse activities may be examined in terms of five types of strategic response to displacement—maintenance, cooperation, contestation, subversion, and withdrawal—depending upon the particular form they take and the impulses that motivate them.

THE PROBLEM OF ACCESS

In the old days the mayor and erstwhile businessman Ivan Allen had a standing lunch date each week with his friend Richard Rich, the head of Rich's department store and one of the most influential men in Atlanta. Mayor Allen also belonged to the Commerce Club, which had been founded by the powerful banker, Mills Lane, as a downtown gathering place for Atlanta's business elite. In that period the world of the city's political and nonpolitical elites was more or less one and the same. To speak of the access of influential business figures to political elites in such a situation scarcely conveys the nature of the relationship between the two groups, for the term implies the ability to pass across a boundary between two realms where roles are clearly differentiated. Such distinctions were blurred in the old Atlanta, however, at least in the white community, rendering irrelevant any concerns among nonpolitical influentials about catching the attention of those who filled formal political roles.

Although such a merger of worlds never occurred in Detroit, in the years before Cavanagh, local businessmen in particular were important to mayors. As Banfield noted in the early 1960s, "[T]he firms that are tied to the city . . . exercise a great deal of influence—aided by the nonpartisan system, for without parties to raise campaign funds, every candidate needs business friends [1965, p. 59]."

Where nonpolitical elites may count on talking to the mayor as a matter of course or, alternatively, where they do not want to maintain such contact, the question of access is almost never problematic. In a number of situations, however, access cannot be assumed nor the benefits of having the mayor's ear ignored. One of these situations occurs with the election of a black mayor. With the elevation to the mayoralty of a spokesman for a constituency never accorded such priority, whose interests, moreover, are potentially in great conflict with those of white elites, access to city hall becomes uncertain.

In Detroit and Atlanta elites therefore sought to pursue a maintaining strategy aimed at the preservation of channels of access that they had enjoyed under previous mayors. Business elites in particular in both cities hoped to trade their early financial support of the candidacies of the black mayors for guaranteed later access. A banker in Atlanta (I-219) explained the strategy: "I'd always been persuaded that Maynard was going to be elected. We did a poll, and it showed he was a winner. So I thought somebody in the business community better get behind this guy so that we'd have a line of communication." To that end the banker and several others of equal stature provided a significant campaign bankroll for Jackson in his contest with Sam Massell. Although a few people involved in this effort suggested that they actually liked Jackson, or liked him better than Massell, the principal reason for providing support did not seem to be for Jackson's political appeal, ideology, or personality, but rather for the access they hoped it would ensure. Several prominent figures in Detroit made the same calculations. The chief fund-raiser for Young, a downtown businessman (I-136), put it this way:

I was able to get the business community involved [in the campaign]. A lot of people gave to both candidates. But I was convinced a black man would win. I had an easy commodity to sell. Businessmen are pragmatic creatures. I told them they'd better be there at the beginning.

Whether the provision of campaign support by white businessmen made the difference between an open and closed administration as far as the demands and influence of the white community were concerned is uncertain. The giving of such support, however, demonstrably fueled white expectations of access, which were by and large not disappointed in either city. Detroit elites almost universally believed that they personally had easy access to the mayor or that their institutions—unions, corporations, consortiums, business organizations—did. Although a few people complained about the occasional tendency of Young's staff to protect the mayor from callers, a problem they attributed to inexperience, the general feeling, as one labor leader (I-14) put it, was that all one had to do to get through to the mayor was to "pick up the phone."

The situation in Atlanta was more complex. Major leaders of the business sector—the activists in the Chamber of Commerce and Central Atlanta Progress, the major developers and bankers—all claimed to have adequate access to Maynard Jackson. Yet among journalists and politicians particularly, the view prevailed that business elites had been shut out of city hall. "His real failure," a city councilman (I-223) said of Jackson, "is his lack of accessibility to the downtown types." A city hall reporter (I-25) commented, "The old bunch hoped they could get along with Maynard, but that collapsed in a couple of months. These guys can't get through to city hall now. Before Maynard, *they* were the ones who were called." Certainly white elites agreed in general that the influence of the business sector had diminished under Jackson; but in fact those who wished to see the mayor were apparently able to do so.

That the perception of an administration closed to prominent whites persisted, despite personal testimony to the contrary, is powerful evidence of the intractability of early communal apprehensions about a black mayor among whites.[1] However, one explanation for the failure of the perception of the mayor's accessibility to conform to experience may

---

[1]Some opportunities for white elite access to the mayor were even institutionalized in Atlanta. Aside from various regular breakfast meetings, the most notable effort involved Action Forum, an organization formed in 1969, consisting of 12 white and 12 black business leaders. During the first term at least of Jackson's mayoralty, the group gathered on Saturday mornings (to emphasize its informality and maximize its privacy) to discuss the state of the city and race relations. Membership was by invitation only, and its sessions were neither open nor reported to the press. Mayor Jackson met periodically with Action Forum and its subcommittees (I-218).

be that those who did gain access believed that such contacts were relatively fruitless. Business elites said that Jackson was either unwilling to respond to their demands and needs, even when these were communicated (I-215, 219, 222, 216), or that the mayor and his supplicants did not "speak the same language" (I-210, 211).

The evidence regarding the approach to the problem of access indicates that elites in both cities were successful in keeping open channels of communication with city hall. If Atlanta businessmen and the mayor were often felt to talk past one another, at least the tradition of communication that had always prevailed survived transition. That white elites sought access to their mayors—that is, that they pursued a maintaining strategy in this area—and that they worried about access suggests three related conclusions. One is that the mayor in particular and city government in general were seen as important enough to warrant attempts to influence them. Second, the fact that elites sought access in order to have their voices heard indicates that in certain areas at least these elites could not satisfactorily meet goals they saw as desirable through autonomous efforts, that is, without government action or acquiescence. Finally, the effort to maintain access indicates that significant white elites did not give up or withdraw from local affairs, leaving the conduct of municipal government entirely to the new majority.

INDIVIDUAL POLITICAL ROLES: FINDING A NICHE IN THE NEW POLITICAL ORDER

A major effect of transition is to bring about a shift in the relative value of vantage points from which an individual may hope to exercise influence on government. Vantage points that provided easy leverage lose their power as the old political center is moved, while points, once on the fringe of the political order are now located closer to the center. In black-mayor cities, to give one example, the vantage point offered by leadership in the black church community rises in value as that offered by leadership in the white business sector declines. Since individuals are by and large bound to one vantage point by virtue of their roles, transition has the effect of bringing new people within striking distance of the center of power and removing others. Naturally, adjustment to one's new vantage point is necessary, whether it has become advantageous or disadvantageous. Shifts in the relative proximity to power of these various vantage points serve simply to redistribute bias in the political order. Individual initiatives are required to capitalize on one's newfound advantage or to overcome new obstacles to the exercise of influence.

Most obviously, as with any change of regime, the constellation of intimates who make up the mayor's inner circle undergoes complete transformation. Unlike ordinary electoral turnover—in which essentially partisan considerations are crucial—the transition from white to black rule means that race becomes an important credential for admission to the inner circle. Such a racial test may not, of course, be applied absolutely, but only certain sorts of people are likely to be exempted. Various types of political mavericks and younger people who were not established figures under a white mayor are commonly candidates for exemption. If such people manage to gain the confidence of the black mayor, they are likely to become interpreters of interests and attitudes of the white community. Just as certain figures had once been identified by whites in the old order as "spokesmen" for the black community, so the whites admitted to the mayor's inner circle come to be regarded by blacks in the new order as ambassadors from their racial community. In general such spokespersons or ambassadors are likely to be chosen by the dominant group rather than by those whose lives and interests they interpret and represent. But such individuals must nevertheless present themselves or be available to play such a role. Thus one individual political adjustment to transition, cooperative in character, is to seek access to the new inner circle.

White elites in both cities agreed that their respective mayors placed their greatest trust in a small number of black advisors, but each mayor was also seen to rely to a limited extent on several whites. During Jackson's first term, Jules Sugarman, whom the mayor had brought from New York to be the city's chief administrative officer, was thought within Atlanta's city hall to be the mayor's closest white advisor. Several young white businessmen and lawyers, none of whom were regarded by the city's major white elites as true members of the "establishment," were identified as the whites outside of government who could most readily claim the mayor's confidence. "Jackson's white allies," a prominent lawyer and former elected official (I-22) noted, "include some of the younger businessmen. None are in the elite. . . . They're successful but not Chamber of Commerce."

Sugarman, widely characterized as an outsider, and several of the younger men, thought to be "naive," clearly fit the description of the types most likely to gain access to the black mayor's inner circle: One had been an activist in the Georgia Republican party long before it was a going concern; another owned a modest business on the outskirts of the city; a third had a reputation as an abrasive political manager. All were young and none were identified as central figures in the constellation of major white elites.

Although Coleman Young maintained more openly cordial relations with prominent white elites than the mayor of Atlanta did, the pattern of admission to the inner circle in Detroit resembled that in Atlanta. The whites upon whom Young relied most were, as in the southern city, young, formerly on the fringes of power, and occasionally from outside the city. The mayor's executive assistant, for example, was a young white lawyer who had worked as a legislative aide to Young when he was the leader of the Michigan Senate. Another man, a labor lawyer, having contributed a substantial amount of money to Young's campaign, asked the new mayor to appoint him to the Police Commission. He and several other whites on the mayor's staff or among his nongovernmental advisors were characterized by one Democratic party official (I-115) as "old left wingers and radical chic types. . . . These are his own coterie, not people with whom he simply has an alliance like the union movement."

The availability of such opportunities for whites is to some extent evidence of the black mayors' needs for certain white resources. That the mayors were able to enlist people to play the comparatively intimate roles of ambassador, interpreter, specialist, or aide also indicates that there were some, particularly those who were not members of the formerly established elite, who were willing to lend their resources to the black mayor. By doing so they were pursuing a cooperative strategy, fully acknowledging the mayor's leadership and political primacy. In addition, both Young and Jackson were able to draw heavily on whites to fill department and agency headships, and board, task force, and commission positions that carried with them no promise of the intimacy of the mayor's coterie. The two executives publicly pursued racially balanced appointment policies with great success (see Table 7.1 in Chapter 7).

Whites in both cities were able to contemplate other strategies for individual influence in local politics and government besides membership in the mayor's inner circle or in his administration. One of the more common of these was to participate in the financing of the black mayors' election campaigns. Given the relatively greater wealth of the white community compared to that of the black community and the need of black candidates for campaign money, the roles of fund-raiser and donor apparently remained as open to whites under black political dominance as when whites monopolized mayoral competition.

Whites' efforts to fill these roles in the two cities may be viewed as both cooperative and maintaining strategies. In Detroit it was common knowledge that Henry Ford II gave $3000 each to Young, Ravitz, and Nichols during the primary campaign in 1973. More significantly, the organizers of the entire fund-raising efforts for both Young and Jackson were white, and, according to the two mayors' respective campaign finance chairmen (I-15, 231), so were the major donors. Although the people who managed the financial aspects of the mayors' campaigns included several whom we have classified as members of the mayors' inner circles, there were also several figures in each city from the established political or business elite who were actively involved. Neither these latter men nor the major donors—bankers, real estate developers, builders, auto company executives—went on to seek the level of intimacy with their respective mayors that membership in the inner circle carried with it. A few gave their time and money out of a political commitment to their candidates; others supplied money in the hopes of guaranteeing access and attention after the election. But whether the motives were cooperation or maintenance, financial resources from the white community largely provided the wherewithal to run the black mayors' campaigns.

Transition may be seen to have enhanced the value of the vantage points in the political order of certain whites formerly on the fringes of power who managed under a black to gain access to the inner circle. But it did not seem to affect the position of money suppliers, who went on doing what they had always done. The shift to black power did seriously devalue other vantage points, however, and those who operated from these positions adjusted primarily through various withdrawal strategies.

For example, several men who would have been situated advantageously under a white majority to compete for the mayoralty were compelled by transition to reconsider or redirect their political ambitions. In both cities the presidency of the city council (the vice mayoral position in Atlanta prior to the new charter) had previously carried with it virtually automatic entry in upcoming mayoral contests. Under black rule the opportunities inherent in this position seemed diminished. Thus after 2 years of testing the political waters, the president of the Atlanta city council, Wyche Fowler, often mentioned as a potential mayoral candidate, redirected his political ambitions to compete successfully for the congressional seat opened by Andrew Young's appointment to the United Nations. Detroit's council president Carl Levin dismissed thoughts of the mayoralty on the grounds that a white probably could not win. He eventually went on to win a seat in the U.S. Senate in 1978. Another man, who might under other circumstances have been heir apparent to the mayoralty as the son of former Mayor Allen and as president of the Chamber of Commerce, denied interest in elective politics, despite encouragement from some Chamber activists, in order to concentrate on his business and civic interests. Yet another formerly prominent figure in Atlanta politics, a southern liberal whose mayoral ambitions had once been supported by the Chamber "crowd" (I-22) commented: "I'm dealt out, . . . I'm too white for the blacks. I couldn't win any office I'd want."

Although those whites who once would have counted themselves among the potential contenders for the mayoralty by and large withdrew from mayoral competition after transition, they nevertheless maintained an involvement in civic affairs or politics at other levels. For another group, however, transition to black rule represented the coup de grace to a thorough withdrawal from local affairs that was already well under way by 1973. Those who had been known in Atlanta parlance as the "shakers and movers" of the 1960s had lost influence with the election of Sam Massell, "the politician who kicked the Atlanta Chamber of Commerce out of City Hall [AC, 27 March 1975]." Members of this amorphous, but nevertheless exclusive, group, led by former Mayor Ivan Allen, had thrown their support behind Charles Weltner, a lawyer and ex-congressman, in the 1973 mayoral primary. But this losing effort was merely a final gesture. Decimated by the retirement of banker Mills Lane, the departure from the city of Constitution editor Eugene Patterson, and the deaths of Richard Rich and Ralph McGill, the group lost its sense of critical mass and energy. Aging Robert Woodruff of Coca-Cola, the city's chief philanthropist, receded into the background, and Allen himself foreswore public involvement at the local level. Although age and the rise of a new generation of leadership made the retirement from the scene of these leaders of the 1960s inevitable, the election of a black mayor effectively cut the city's links to the particular past these survivors represented.

Some of the individual political adjustments made by whites had the effect of providing the black mayor with the resources of expertise, experience, money, and putative links to the larger white community. This was particularly true in the cases of those young men who sought and gained admission to the mayors' inner circles. Their cooperation was of much the same character as that of the young Mugwumps in Boston, who acknowledged and served Irish rule in the early years of transition. Other adjustments, however, although they added no stock to the treasury of resources the black mayor could command, at least cleared the field of potentially troublesome and highly visible competitors. The redirection of political energies by some possible mayoral candidates and the reconsideration of political careers by others left the mayors free from concerted challenges in the early years of their terms. Only in the case of the withdrawal of Atlanta's old guard can it be said that significant resources were withheld from the occupant of city hall, but the withdrawal of these figures was in fact already in progress when Maynard Jackson was elected. A white mayor chosen by the old guard could have done no more perhaps than slow their inevitable departure from the scene.

THE QUESTION OF ELECTORAL CONTESTATION

One of the most striking consequences of transition, at least in the short run, was the attenuation and disorganization of potential opposition to the new regime. Displacement could, of course, have led alternatively to the hardening of opposition, as the new minority sought to buttress itself to withstand the loss of power. But instead, as we have seen, the normal pool of white candidates had virtually evaporated halfway through the first terms of the mayors. Moreover, although some of the reluctance to declare an interest in the mayoralty at such an early stage could in other circumstances simply be construed as a matter of political coyness, in Detroit and Atlanta there was also a lack of organized effort to lay a political base for any of these men, which lent a degree of truth to their disavowals. Indeed, the normal pool was not reconstituted for the 1977 elections in either city. Thus, both individual ambition in the service of opposition to the black mayor as well as collective efforts to encourage and support such ambition were lacking in the two cities. Laying a basis for electoral contestation was, therefore, not a major strategy for white elites in either place.

There was no lack of either speculation about possible candidates or wishful thinking. But all speculation was structured, however, by the widely held assumption among white elites in both cities that the personal political strength of both incumbent mayors was so high as virtually to ensure them second terms. Not only were demographic factors acknowledged as favoring blacks, as we have seen, but no sources of opposition capable of leading an electoral challenge could be identified. White homeowners and police in Detroit and the newspapers in Atlanta were perceived as the most vocal opponents of the mayors, but none of these was thought to possess the strength or means to mount a challenge.

Among those in Atlanta who did control the resources necessary to groom and back a serious candidate, there was neither agreement on who such a figure might be nor confidence in the capacity to win of any of the people mentioned. One man known for his ability to identify and manage successful candidates summarized opinion regarding the two most commonly mentioned names: "Allen doesn't have a ghost of a chance. Fowler couldn't win either [I-232]." Although some among the white business and political community wished for a white "star" to run against Jackson, others were willing to consider the possibility of a black candidate more congenial to white interests. Yet no suitable candidates had been identified. One major construction executive (I-215) commented: "People are looking for a qualified candidate of either race who will agree to run. That's the toughest part: There's no one to bet on."

Elites in both cities agreed that black control of city hall had produced an artificial, but nevertheless strong, unity in the respective black communities. Ambitious black politicians seemed inclined to bide their time, deferring to the mayor. "Within the black community," a Democratic party leader in Detroit (I-115) commented, "there are many with some disaffection from Young. But it's kept inside. There's an agreement not to criticize him." Nevertheless, this tacit deference did not entirely prevent the floating of names of several potential black challengers in each city. These possible candidates were not, however, taken very seriously, although significant numbers of white elites, particularly in Detroit, saw these figures as attractive alternatives to the incumbents.

To summarize, preparations by whites to engage in contestation were virtually nonexistent. Not only was there no publicly chosen candidate whom significant white elites sought to groom or encourage, but there was no hint in either city of any sort of clandestine effort to produce a candidate. No money was being raised, no groups were being formed, and no strategy had been established. And finally, there were no self-declared white mayoral hopefuls who would admit their ambition in public or private. In short, not even the buds of electoral opposition had emerged.[2]

Perhaps the formlessness of electoral opposition strategies or their general lack altogether is not uncommon in American cities at the midpoints of mayors' terms. Planning to campaign for city hall for 2 years may simply be regarded as a gross overexpenditure of effort. But whether or not the failure to generate early efforts to mount an eventual challenge to the incumbent mayor conforms to the experience in other cities, this inactivity must nonetheless be taken as a significant measure of the extent to which transition was essentially accepted in those places. This does not mean that serious white electoral opposition cannot develop in these cities, nor that serious efforts to find "manageable" black candidates will not occur. But what is important is that displacement in Atlanta and Detroit did not stimulate a cohesive effort to organize an electoral challenge as an early strategic response.

[2]The reliability of these findings is borne out in the elections of 1977. White challengers in both cities emerged late and came from the fringes of political power. None were prominent "establishment" figures and none had significant backing from major influentials in the white communities. The only serious challenger in either city was a black city councilman in Detroit, Ernie Browne, who attacked Young from the right.

In part this may be explained by the fact that certain elites, particularly in Detroit, supported the black mayor and wished to see him succeed. But the more important explanatory factor in both cities was perhaps the demographic balance, which now favored blacks. Although there was no evidence that the change to minority status demoralized white elites in either place, it is evident that some of them redirected or reconsidered personal political ambitions. The absence of obvious white candidates of star quality had the effect of both weakening and diffusing the focus of those who traditionally organized and funded mayoral candidacies. Thus no strong consensus emerged on whom to groom or support.

THE ADJUSTMENT OF MAJOR ECONOMIC ACTORS

In cities with black voting majorities, white "economic notables," as Robert Dahl once called them, are not wont to claim that they control or even influence all matters of civic importance. Even if they once believed they did—as in Atlanta in the Hartsfield–Allen years (Allen, Jr., 1971, pp. 29–31, 240)[3]—the very existence of a black mayor makes the notion of a cohesive elite of white businessmen who monopolize the apex of the local power structure no longer tenable. Yet the power of a city's principal economic actors, if neither unified nor all-determining, is nonetheless undeniable. Through a host of generally uncoordinated private decisions based primarily on market factors, a city's businesses and banks and developers exercise a major impact on the quality of urban life.[4] To the city's governors the decisions of a bank to buy or forego a city's notes, of a corporation to keep its headquarters downtown or to move, or of a developer to build a shopping center in the city or on the metropolitan periphery all bear ultimately on the prospects for successful government. The availability of credit, the production of jobs and tax revenues, and the ability to lend a city a reputation for economic vitality through development activities are factors controlled largely by private economic actors and upon which the fortunes of municipal government often depend.

To suggest, however, that corporate and commercial economic actors make their impact felt only through spending and investment deci-

[3]In contrast Detroit's elites from both the corporate and labor sectors agreed that the major corporations—especially the auto makers—never "ran" the town. Although their influence has always been felt, the United Auto Workers, as well as nonautomotive industries, have exercised a degree of countervailing force. In addition, mayors have never been regarded as representatives of the auto industry (I-115, 114, 119, 131, 135, 123, 137).

[4]See, for example, the arguments of Warner (1968) and Long (1958).

sions based on market considerations—that is, profit and loss factors—is not entirely accurate. Particularly since the civil disorders of the 1960s, urban-based firms and banks have developed a sharper sense of civic responsibility (Neubeck, 1974, p. 5). This was manifested in the late 1960s in part by a surge of corporate activity in social action programs. For example, one study of 247 of *Fortune's* list of the largest corporations in the United States showed that by 1970, 201 had established urban affairs programs, only 4 of which predated 1965. One hundred and seventy-five of these firms revised their annual donation lists after 1967 to provide grants to groups specifically identified with urban problems. One in four reported donating staff and facilities to antipoverty agencies, Model Cities projects, community groups, or the local urban coalition (Cohn, 1971, pp. 3–20).

That corporations might respond in these ways to urban rioting in the very streets in which their offices and plants are located is understandable. As Cohn (p. 4) notes, a major motivation for spending resources on "urban crisis" programs has been not simply to meet certain vaguely defined civic responsibilities owed to the place in which business is done but also to create the social stability and skilled work force that make for a healthy business climate. But the stimulus to this new style philanthropy—imminent mass street violence—is long since past in Detroit and Atlanta. Black mayors and black majorities in and of themselves do not threaten social stability, at least not in the sense looting and burning do. Thus the question to ask is whether commitments to the health of the city, through both private investment and development and support of nonprofit urban programs, have been affected one way or the other by transition to black rule. Were corporate actors, by their support of urban social programs and continued investment, seeking to stabilize or revive the urban environment only as long as it was ruled by whites?

The question of business commitments cannot, of course, be understood simply in terms of whether private investment or participation in social welfare and urban affairs activities took place. The issue of how much participation or how much investment is equally important. Yet making valid intercorporation comparisons or setting standards of participation raises essentially insoluble difficulties. Businesses have different capacities to demonstrate their commitment, a function in part of different profit margins and capital liquidity. They also apparently have a variety of equally valid ways of addressing their civic obligations. Providing a monetary grant or the loan of an executive to the local urban coalition may offer a satisfactory way for one corporation to demonstrate its sense of civic obligation, whereas developing a low- and middle-income housing project in the city may represent an equally important but noncomparable way of fulfilling another firm's commitment. A banker in Detroit (I-135) summed up the whole dilemma:

> You realize how hard it is for an organization which is part of the free enterprise system to quantify what it ought to be doing in the socio-economic/social welfare area. We budget a minority lending division in the bank for a loss of $250,000. We want to be good corporate citizens, but what does that mean? What standards do you bring to bear?

Given these problems, the question is whether or not there was evidence after the transition to black rule of commitment to Atlanta and Detroit on the part of their major economic actors.

The behavior of major economic actors, both historically and under the black mayors, has been substantially different in the two cities. Atlanta, for example, enjoyed a tradition for many years of individual private philanthropy for public purposes. Led by such people as Robert Woodruff of Coca-Cola, Atlanta donors supplied funds from their personal fortunes for cultural facilities, public parks, and Community Chest programs. This tradition of individual giving was matched by the entrepreneurial efforts of individual bankers and developers who supplied the energy and resources to build Atlanta's Civic Center, sports stadium, and some low-income housing.

During the Hartsfield–Allen years the collective efforts of the business community were channeled through the powerful Chamber of Commerce. As the power of the Chamber began to wane after the election of Massell, business influence came to be exercised through Central Atlanta Progress, a privately funded organization designed to encourage investment in the downtown. "We are devoted," its director (I-216) explained, "to putting business and civic power into things the city should be doing."

By comparison the record of Detroit's economic leadership was one of indifference prior to 1967. With the exception perhaps of Henry Ford II (Conot, 1974, p. 444), Detroit has no history of individual patronage during the boom years of automobile production. A former city administrator (I-15) explained:

> The fat cats of Detroit have never really done much for the city in comparison with those of other cities. They didn't give a damn until the riots. They dominated economically but they had no attacks of conscience. . . . Traditional elites in the economic world here have been corporate elites rather than family elites, especially after World War II.

The riot of 1967 marked a turning point in the attitudes of Detroit's major economic actors (I-18, 125, 133, 135). In contrast to Atlanta, however, collective energies were not initially organized through a promotional organization devoted mainly to real estate development but rather through an urban coalition called the New Detroit Committee, whose principal focus has been on the stimulation and support of job and social programs.

By all indications the commitment to Atlanta and Detroit of major economic actors, both through the ongoing generation of profit-making activities and participation in public service enterprises, continued under black rule. If, as in some but not all cases, the investment of resources by individual firms in either of these areas diminished during the black mayors' first terms, the perception of the economic elites themselves was that the causes could not be attributed to the rise of black rule but rather to national economic pressures. There is little evidence to doubt this interpretation. The following examination of particular economic organizations will serve to confirm and enlarge upon this conclusion.

CENTRAL ATLANTA PROGRESS

By the time Maynard Jackson was elected mayor, Central Atlanta Progress (CAP), formed in 1967, had become the major vehicle of business influence in the city, surpassing the Chamber of Commerce. With a membership heavily composed of downtown merchants, real estate developers, and financial institutions, plus the city's major corporations, its main interests was, as one of its former presidents (I-228) put it, "to keep central Atlanta alive." CAP's activities, primarily research, promotion, and public relations, were financed by members' dues and by occasional earmarked grants from the federal and city governments. Its annual budget was approximately $250,000.

Several aspects of the organization are important for our purposes, for they illustrate how Atlanta's economic elites maintained an interest in their city, yet in a wholly characteristic way. For example, Central Atlanta Progress did not for the most part pursue public service or social service goals; its purpose was to enhance the downtown as a marketplace, a place in which to do business. Thus a number of its projects, many of which had public good ramifications, involved the development of plans to attract people (potential customers) to the downtown. CAP pushed for pedestrian malls and improvements in traffic circulation patterns, and it provided some funding for cultural programs in central city parks and the decoration of blank building walls with giant murals. It encouraged the spread of high-intensity street lighting and additional police foot patrols to combat crime. For downtown businesses it established a "war room" to provide up-to-the-minute information on the nature and schedule of disruptions caused by the construction of a new underground mass transit system. CAP was also active in seeking to attract business to the downtown area and in encouraging its member lending institutions to finance home mortgages in the inner city to draw middle-class families back to the center. The only major activity of the organization that did not have implications for enhancing the prosperity of central city businesses was a project undertaken in cooperation with the public schools to match job opportunities with vocational education training.[5]

A second feature of the organization that is important was its vision of itself as a cooperative partner with city government. Writing to Maynard Jackson during his first year in office, the president of CAP, Harold Brockey (16 Sept. 1975) spoke of the need to "reforge the progressive partnership between business and City Hall centering on the commitment to the viability of Downtown." Many of CAP's particular projects were carried out in conjunction with or recommended to public agencies and departments such as those responsible for police, schools, parks, mass transit, and housing. Thus CAP sought to maintain a cooperative relationship with city government even after transition, carefully avoiding any appearance of an adversary posture.

A third feature of CAP was its willingness to generate profit-making development activities in the inner city. Shortly after Jackson's election, CAP organized a consortium among its members, called Park Central Communities, to undertake a $250 million housing and commercial development of a largely vacant 78-acre urban renewal tract in the downtown. The plan was conceived and developed with the cooperation and encouragement of the mayor, the Atlanta Housing Authority, and neighborhood groups. Oriented heavily toward middle- and upper-income housing, the development was designed to be a profit-making venture. At the same time, it represented a statement of business faith in the essential economic health of the city.

The activities of Central Atlanta Progress suggest not so much a series of desperate lifesaving measures designed to salvage what one might from a foundering business district but rather a variety of attempts to preserve and enhance a basically good business environment. The emergence of a black majority and the election of a black mayor did not dampen basic entrepreneurial impulses. CAP strategies toward city gov-

---

[5]The summary of CAP projects is taken from Central Atlanta Progress (1975).

ernment were a combination of maintenance and cooperation, promising benefits to both its member businesses and to the city generally. As if to emphasize its continued commitment, the CAP executive committee met in August 1974 "to document or dispel" rumors that businesses were fleeing the downtown "for other than economic or management reasons [Brockey letter]." The committee decided that there was no discernible trend indicating business flight, but it went on to offer a set of proposals to the mayor that it believed would ensure the continued health of the central business district. The local press interpreted the "Brockey letter" as a highly critical attack on the mayor, but neither Jackson nor CAP understood the communication in those terms. For both the senders and the recipient, the actions of CAP were evidence of the commitment of Atlanta's major economic forces to the survival of the city as the economic center of the metropolis.[6]

DETROIT RENAISSANCE

The organization in Detroit most comparable to Central Atlanta Progress is Detroit Renaissance, a nonprofit corporation founded in 1970 for the purpose of identifying construction and expansion opportunities in the downtown and attracting investors. Its broad aim is the "brick-and-mortar" revitalization of the central business district in the pursuit of a greater tax base and increased employment and sales. Like Central Atlanta Progress, it is funded mainly by assessing its membership, which is composed of a restricted number of major firms in the Detroit metropolitan area. Its budget in the early 1970s came to approximately $750,000 per year.

Detroit Renaissance engages in public relations, housing feasibility studies, and the funding of wall murals on building exteriors, but its major efforts are devoted to the encouragement of investment. Its director readily took credit for the organization for stimulating a substantial increase in office construction in the central city between 1972 and 1976. In those years (virtually the years of Young's first term) over 5 million square feet of office space were completed, compared to slightly more than 2 million square feet built between 1961 and 1971 (Detroit Renaissance, n.d., p. 2).

Detroit Renaissance is identified as having played a key role in aiding the formation of the largest private investor group ever to undertake a redevelopment project, the Renaissance Center Partnership (I-124).

---

[6]For the mayor's views on the Brockey letter, see the interview with Maynard Jackson, "Can Atlanta Succeed?" (1975, p. 110); see also Powledge (1975, p. 46).

Goaded by Detroit Renaissance president Robert McCabe to show what "business was going to do about Detroit," Henry Ford II assembled a consortium of 51 corporate investors during 1971–1972 to erect a one-half billion dollar office, hotel, and apartment complex on the city's riverfront (Stark, 1973, p. 6). The Center was partially opened in 1977 with nationwide media fanfare. Downtown businesses reported an almost immediate surge in retail sales (NYT, 24 July 1977). Some Detroit figures had nevertheless predicted that the fortress-like structure would simply drain existing office space of its wealthier occupants (I-133, 136). Others, however, saw it as a major symbol of the rebirth of the city and a demonstration of business faith in its future (I-110, 118, 124, 135).

NEW DETROIT

Detroit's violent episode in the long hot summer of 1967 ended during the morning hours of August 27. That afternoon Governor George Romney and Mayor Jerome Cavanagh invited 500 selected Detroiters to meet with them in the City–County Building to consider the future of the city. The gathering included representatives of neighborhood and minority organizations as well as the heads of the major automobile companies and labor unions. From this meeting the New Detroit Committee was formed, the first "urban coalition" in the nation. A year later it was incorporated as a nonprofit organization. It then hired a professional staff and commenced full-time operations. It has no counterpart in Atlanta.

The idea of an urban coalition—later copied in several dozen other cities as well as at the national level—was to consolidate and bring to bear the varied resources of the private sector for public purposes and to provide a forum in which government, business, labor, and citizen leaders could regularly talk with one another. Funded entirely by grants from private corporations, labor unions, and foundations, New Detroit was able to operate on annual budgets that never fell below $2 million.[7] Nearly every major firm in Detroit sent its highest ranking official to sit on the board of the organization. The board also included representatives of labor, minority groups, and churches. Although the president of New Detroit was a black man throughout Young's first term, the chairmanship was generally reserved each year for major white business figures: Richard Gerstenberg of General Motors, Joseph Hudson of Hud-

[7]Between 1968 and 1974 New Detroit raised $24 million. Its highest budget (the first year) was $10 million. Thereafter, and all through Young's first term, the budget was set at $2 million per year (New Detroit, 1974–1975, 1975–1976).

son's Department Stores, and Max Fisher, the financier, were among those who held the position.

The aims of New Detroit were similar in a broad sense to those of Detroit Renaissance and Central Atlanta Progress, namely to mobilize and channel private resources to help revitalize the city. But in its particulars New Detroit's strategy was different. Committed to encouraging "social change," New Detroit concentrated on reforming existing public agencies and providing seed money grants for a wide variety of social programs and minority business ventures. The assumption under which Central Atlanta Progress and Detroit Renaissance worked—namely that a healthy marketplace makes for a healthy city through the spillover effects of business growth—was conspicuously absent in New Detroit; although 13 of the 26 board members also sat on the board of Detroit Renaissance, New Detroit concentrated on activities aimed at enhancing economic, health, and social opportunities for poor and minority groups, as illustrated by the following examples of its projects.

1. Financially supported a public relations program to ensure peaceful implementation of a busing plan for Detroit's schools
2. Assisted in the decentralization of the Detroit school system
3. Advocated and helped to implement recruiting and testing reforms in the police department designed to draw in more minority candidates
4. Stimulated the establishment of a new community college
5. Funded a sickle-cell anemia screening program
6. Initiated hiring programs for the hard-core unemployed in private industry

The only activity designed to bolster private sector business capabilities that New Detroit undertook was the establishment and funding of the Inner City Business Improvement Forum (ICBIF). New Detroit provided ICBIF with nearly $4 million between 1968 and 1975, a sum that was more than matched by foundation and federal government grants. The money was used to provide venture capital for minority-owned businesses and to support technical assistance services to those enterprises receiving loans.

Although the organization was not exclusively a vehicle for the white business community, its funding, prestige, and visibility derived almost entirely from white sources. If significant white elites had begun to lose interest in Detroit after the election of a black mayor, such feelings would have been reflected in the operations of New Detroit. If anything, however, the relationship between the city government and

New Detroit was strengthened by Young's election. "Under Gribbs," a prominent figure in the organization (I-117) remarked, "New Detroit was in more of an adversary relationship. . . . We've gotten what we asked for now [in regard to police reforms] under Coleman, but we didn't get it under Gribbs." Not only was Young seen as more sympathetic to the "social change" goals of New Detroit than his predecessor had apparently been, but he had earned a reputation as an honest, competent, flexible member of the New Detroit board during his days in the Michigan Senate (I-139). When Young put together an elaborate $2.8 billion proposal based on federal money (the first "Move Detroit Forward" plan) to be used for a variety of physical development projects, prominent white New Detroit board members accompanied the mayor to Washington to meet with President Ford to win his support. During the first 2 years of Young's mayoralty—the worst years of the economic recession—New Detroit still managed to raise its target funding. "Even when the corporations were losing money," the chairman of New Detroit (I-139) commented, "they were still contributing."

THE CHAMBERS OF COMMERCE

Although Detroit Renaissance and Central Atlanta Progress participate in the promotion of their respective cities as unequaled places to do business, the principal responsibility for such campaigns lies with the local Chambers of Commerce. Neither Chamber was as influential during the 1973–1977 period as in former years. Perhaps the main reason is that Chambers lack the capacity to undertake actual investment ventures and thus have been superseded by consortia of businessmen who can. Newer vehicles, such as Central Atlanta Progress and Detroit Renaissance, emerged to consolidate and coordinate business resources in the pursuit of collective money-making activities. These activist strategies not only promised more direct economic benefits for investors than those that arose from successful Chamber efforts to lure out-of-town firms to relocate, but they seemed also to produce more spectacular publicity for the city. The cases of the Bedford-Pine housing development in Atlanta and Detroit's Renaissance Center illustrate the point.

Nevertheless, both Chambers—especially that in Atlanta—have long traditions of communicating local business opinion to city government and of "selling" their cities to the outside world. After the election of the black mayors both of these general activities continued, cloaked in an official policy of support for the new city administrations. The president of the Atlanta Chamber during Jackson's first year (I-211) commented on the relationship of the organization with the mayor:

> The Chamber has taken the position that he's the mayor, a responsible public official, and we're going to work with him as closely as possible. We resolved that even before he was elected. We're going to help him when we can and tell him when we disagree. It makes good sense for an economic organization like the Chamber.

That the Atlanta Chamber resolved to cooperate with the mayor before he was elected suggests an attempt to frame the relationship in impersonal terms as one between roles in which the personal characteristics of the incumbent of the mayor role are incidental. To the extent that the Chamber was to be bound by the traditional obligations of its role, it sought as far as possible to avoid the appearance of an adversary stance and to cooperate, as it had always done, in the interest of business. It is important to note, then, that the advent of a black mayor did not cause the Chamber to redefine its role obligations. The pattern was similar in Detroit. "During the Gribbs–Austin race [in 1969]," the Detroit Chamber president (I-110) explained, "the Chamber held meetings of business leaders with both candidates, so whoever got elected, we would be supportive. We were sharing our views with them to tell them we'd help them, no matter who was elected. We did that with Young and Nichols too."

To sustain their promises Chamber officials in both cities participated regularly in periodic meetings between the mayor and leaders of the business community. In Atlanta these took the form of unstructured "pound-cake breakfasts," which Chamber representatives did not find entirely satisfactory. "We're having a painful time, but we're making it," said one Chamber spokesman (I-234). Detroit Chamber executives participated in monthly meetings with the mayor at which representatives of New Detroit and Detroit Renaissance were also present.

While the two Chambers were carrying out their customary communication functions with city hall, they were also busy promoting the business virtues of their cities. Both organizations made regular forays to other cities in an effort to attract new business, just as they had done before the black mayors were elected. City grants helped to pay for these trips, and, occasionally, city officials went on them. The most dramatic promotional effort during the first terms of the black mayors was launched by the Atlanta Chamber early in 1977 when it announced the opening of a branch office in New York City to serve as a base for recruiting firms from the Northeast. The New York office was to be one component of a $1 million advertising campaign. The characteristically confident theme of the campaign was "Atlanta: A City Without Limits" (NYT, 11 Feb. 1977).

New Detroit, Detroit Renaissance, Central Atlanta Progress, and the Chambers of Commerce all represent slightly different ways of applying collective business resources to the problem of the survival of the city as a vital commercial and social center. None of these organizations had radically transformed its city in the mid-1970s, nor did any of them expect to do so. Certainly none is capable of such a feat. The limitations of their efforts are clearly evident: Their budgets are too small for the goals they pursue, and the demands on their resources are too many. Measured objectively against the problems of their cities, what such organizations do is simply not enough. Not only are their efforts inadequate in this sense, but some of them are also misdirected—from the point of view of the public sector—by virtue of the fact that, to varying degrees, the self-interest of major economic actors is the ultimate motivation for their public-oriented activities. In addition, this self-interest suggests that market forces must at some point impose limits on the play of social conscience impulses, insofar as they are present at all, although the directors of these organizations do not publicly consider the point at which their collective activities might become a serious losing proposition. Theirs is an ideology of optimism in which the assumption that private action can significantly alleviate problems in the social and economic order and be profitable at the same time is a central plank. The whole assumption rests on the slim and uncertain belief that economic spillover effects will indeed create jobs for those most in need and fill the coffers of local government to permit spending for social welfare.

Nevertheless, this nexus of private sector activity cannot be dismissed as entirely self-serving or meaningless for the city. The investments made through or by these organizations—whether of money, prestige, or personnel—must be seen as trustworthy public statements of faith and commitment to their respective cities. It is important to realize that none of these organizations chose, as they might easily have done, to redefine self-interest in such a way as to lead a withdrawal of white resources from their cities in search of another marketplace once blacks won the mayoralty. By staying where they are and by doing what they do, New Detroit, Detroit Renaissance, CAP, and the local Chambers serve, within the limitations of a business context, as extremely important symbols of collective private sector interest in their cities. As such, these organizations may be seen as major business vehicles in pursuit of what the prime economic actors in these cities conceive to be a basically cooperative strategy in regard to local government.

INDIVIDUAL PRIVATE FIRMS

Assessing the nature of the commitment to the city of major individual corporate entities—manufacturers, banks, utilities, and retail stores —is a difficult undertaking. Evidence tends to be spotty and hard to come by. Pieced together, however, there emerges a general picture of a widespread willingness to remain in the city, to invest there, and to help. At least during the first terms of the black mayors there were no signs of panic or withdrawal among the major corporate actors.

Systematic data concerning the number of departures of firms from the city or the rate of ingress are not compiled by any agency or organization. The morale of the business community leadership, however, seems to be shaped less by the net movement of firms in or out of the city than by a limited number of decisions of highly visible major firms. Thus, acknowledged losses of small firms in the service sector were more than offset in the minds of business observers by Chrysler's renovation of its Mack Avenue plant, the announcement in 1977 that General Motors would move its overseas division with 800 jobs from New York City to Detroit, the expansion of Coca-Cola in Atlanta, or the construction there of the massive Peachtree Center hotel complex. Such decisions were interpreted as the truly critical signs of the state of the city's health, and in both cities such announcements fueled a general optimism.

Among executives in major industries and banks in both cities, professions of determination to stay and expand were legion (I-118, 123, 125, 130, 135, 136, 219, 222, 228). A spokesman for the Burroughs Corporation (I-132), the leader of the nonautomotive industries in Detroit, provided an account of his company's commitment that was not untypical:

There was no need for a world headquarters to stay in Detroit. There was pressure to move us to the east coast and to Southfield [a Detroit suburb]. The riots came right after our decision to stay. There was a lot of soul-searching about staying, but then there was a reaffirmation. We've got an ideal up-town location. And now, I haven't heard anyone around here say, "Oh God, we should have gotten out of here" because of the city's fiscal crisis. The budget cuts won't hurt us that badly. We do 90 percent of our own security anyway, and maybe that's the way it should be. There's no panic at all here.

Business leaders interviewed for this study tended to view their own commitment to stay in the city as typical of others. Certainly none of them—and no officials in the Chambers of Commerce for that matter—were able to identify many firms planning to leave the city.

The picture gained from the elite interviews is substantially corroborated by the results of a survey conducted among the major firms in both cities. Questionnaires inquiring about firm relocation and certain capital investment plans were sent in 1977 to the public relations officers of all 121 firms employing 1000 people or more in the Detroit and Atlanta metropolitan areas. Seventy-two firms (60%) responded to the survey.[8] Results for those firms with major facilities located within the two central cities are presented in Table 5.1.

One-third of the firms in each city admitted to having considered moving out of the central city within the 5 years prior to 1977, but only a tiny fraction said they had actual plans to do so. Most businesses, ranging from industrial enterprises to commercial establishments to white-collar industries, had plans for renovation or expansion of existing central city facilities, and a large majority had actually carried out such activities during the period that covered the black mayors' first terms. The fact that reported firm behavior in Atlanta was remarkably similar to that in Detroit provides some measure of confidence in the reliability of the data. In general, commitment to the respective central cities among large employers appeared strong. The data reveal neither signs of massive flight from the two cities nor cautious retrenchment.

Other evidence tends to confirm the sense of commitment to the city professed by corporate spokespersons. In both places, for example, the major firms loan executives to the city government and to various social action and economic development projects. In Atlanta the Chamber serves as the clearinghouse for a modest program of executive loans, which are made primarily to the school district and to city government agencies for the purpose of advising on budget practices and managerial techniques. New Detroit is both a major recipient of loaned private sector executives and the main conduit of this talent to other organizations, particularly those that advise and assist minority-owned businesses. A utility company executive in Detroit (I-123) commented on such practices: "We've loaned people to New Detroit. A former black city councilman joined our law department and was loaned to New Detroit and became its president for three years." Automobile company officers and bankers told of similar contributions to the executive talent pool (I-118, 128, 129, 135).

---

[8]For additional details concerning the response rate, see Appendix B.

TABLE 5.1
A Portrait of Business Confidence among Central City Firms Employing 100 People or More, 1977

	Percentage answering "yes"[a]	
	Atlanta	Detroit
1. Has your firm ever thought in the last 5 years of moving any major facilities out of the central city?	32	33
2. Does your firm have any actual plans to move any major facilities out of the city in the next 5 years?	9	4
3. Does your firm have any plans to renovate existing central city facilities?	41	68
4. Does your firm have any plans to expand central city facilities, including new construction, over the next 5 years?	26	45
5. Has your firm actually renovated or expanded central city facilities in the last 5 years?	80	77
6. Has your firm actually moved any facilities out of the central city to other locations in the last 5 years?	30	36

[a]N's range from 22 to 26, depending on the question, in Atlanta. In Detroit N's range from 21 to 26.

In Detroit the role of the banks in the affairs of the city became especially important during the recession of the mid-1970s as the city confronted the possibility of fiscal collapse. Detroit faced a budget deficit in 1976 of $64 million. The only device open to the city under state law to finance such a deficit was to borrow against the following year's tax levy by the selling of tax anticipation notes. After reviewing the city's affairs, the state Municipal Finance Commission allowed Detroit to offer $40 million in such notes, which, combined with some state aid and local public employee layoffs, would enable the city to balance its budget. Determined to avoid the charges of irresponsibility leveled at New York banks during that city's long fiscal struggle, Detroit banks promptly formed a syndicate, which bid on $27 million worth of notes. "We will bid on the other $13 million," the president of a major bank (I-135) promised. "We held off to push the mayor on the budget cuts (although he's been pretty realistic) and the governor. They both responded." Conversely, the actions of the banking community pleased the administration in city hall. An aide to Mayor Young (I-131) commented:

> I saw no hesitancy on the part of the bankers here. We had no doubt that they'd put together the syndicate and come through. They have too much at stake in the city. There wasn't much reticence really. No one put a gun to their heads. They could see that the mayor was doing the best he could. We really have pretty good support from the business community.

In an effort to follow up on the willingness of the banks to come to the city's aid, Mayor Young formed an Economic Growth Council late in 1976 composed of 56 of the top former and current business leaders in the city. Its purpose was to review Detroit's tax structure, evaluate the performance of the city government, and propose programs to develop jobs. The mayor asked retired chairmen of General Motors (James Roche) and Chrysler (Lynn Townsend) to head the committee. Robert Surdam, president of the National Bank of Detroit, assumed the position of operating vice chairman. Each of them pledged extensive personal involvement on the panel (NYT, 27 Oct. 1976).

Detroit banks were also involved in a well-publicized effort to reduce the practice of red-lining in mortgage financing. The seeds were planted by the city's largest bank when it took out full-page advertisements in the local newspapers in 1976 pledging to loan to any qualified buyer, regardless of the location of the property. Then in 1977 most of the banks and savings and loan associations in the city, working with city officials in the mayor's office, formulated the "Detroit Mortgage

Plan." Under the plan loans would be dependent upon the applicant's credit status and the condition of the block rather than the neighborhood or area in which the property was situated. An impartial panel would review rejected applications. Detroit political leaders and bankers expressed hope that the plan would serve as a model for other cities in the effort to stem movement out of the city and to improve existing commercial and residential stock (*NYT*, 21 Feb. 1977).

These various instances of business mobilization and professions of business commitment do not provide a systematic or even necessarily representative picture of the larger pattern of local economic decisions that affect the two cities.[9] Nevertheless, they suggest that the major economic actors, at least, did not contemplate withdrawal of their resources from these cities and in many cases made certain resources directly available to city government and the mayor in particular. The firms involved in these decisions are regarded as the corporate and banking leaders in their cities. They establish, therefore, the dominant models for business–government interaction, and they create a highly visible symbolic context by which the health of the city is judged by others. It is important, of course, to understand that not all of what these firms do is merely symbolic: Many of their efforts require genuine investments of resources. A few of these offer little return in the form of stockholder dividends. Whether these investments are "enough," whether they are made for the "right" purposes, or whether they do indeed have the intended spillover effects is at best arguable. But it should be stressed once again that within the terms of corporate capitalism, these major firms have chosen by and large to fashion symbols of commitment and cooperation in their respective cities rather than symbols of withdrawal and despair.

THE ROLE OF STATE GOVERNMENT

During the transition to Irish rule in Boston, embattled Yankees periodically called upon the state of Massachusetts to act in ways that would limit the power and prospects of Irish politicians. In response to these demands the Massachusetts legislature, among other actions, re-

[9]Total assessed property valuations show a small but steady decline in Detroit during Young's first term from $4.37 billion in 1974 to $4.06 billion in 1977. The decline is largely a function of small business and housing relocations and abandonments. Similar figures for Atlanta show a steady increase during the same period. Full data are presented in Appendix C.

moved control of the police from the hands of local government, introduced nonpartisan elections, did away with aldermanic districts, assumed oversight of the city's fiscal affairs, and debated various annexation schemes. The possibility that mainly white, suburban, and rural dominated legislatures would respond similarly to white minorities in black-ruled cities—or indeed abandon the latter altogether—was a matter of serious speculation by contemporary observers of urban areas.

During the early years of black control in Atlanta and Detroit there was no evidence, however, to suggest either that whites within those cities viewed the state as a source of relief from their minority status or that Georgia or Michigan sought to restrict black power or abandon their major cities. It is true that a substantial number of Atlanta elites were active in the efforts to effect metropolitanization through state legislative action. But their motives were mixed, as we shall see in Chapter 6. A purely racial explanation for this reform activity does not suffice. Besides, their efforts met with little response in the state capitol, suggesting that the Georgia legislature was in no hurry to "rescue" the city from black rule by turning it over to the white suburban electorate. Aside from its role in the metropolitanization controversy, the state's relationship to the city of Atlanta was not a matter of great concern, interest, or expectation among local elites. They agreed generally that Georgia had little interest in constraining the power of Atlanta local government or in abandoning the city, the shining symbol of the economic rebirth of the South (I-22, 24, 217).

Although the metropolitan reform directed at Detroit has long been a minor issue in Michigan state politics, the critical element here in the state–city relations—and the one on which the commitment of the state might be judged—has been the city's need for more money. White elites were prominent in joining Mayor Young in his appeals for increased state aid during the city's fiscal crisis. For example, in a speech in which he claimed that Detroit was one of America's best managed cities, Chrysler Board chairman John Riccardo urged state legislative and popular support for a plan formulated by the mayor and the governor to supply the city with more money (DFP, 25 April 1976). The plan called for the state to pick up major portions of the cost of Detroit General Hospital, Detroit River patrols, the Art Institute, library, and historical museum, thereby freeing nearly $26 million for the city to spend elsewhere. Executives of other auto companies and major nonautomotive industries echoed Riccardo's call.

White elites viewed both the governor (Milliken, a Republican) and the legislature as committed to the health and survival of the city (I-114,

129, 131, 132, 135). Elite explanations for the amicable nature of the state's relationship with the city ranged from the growing out-state recognition of the interdependence of the state and its major city to the fact that Detroit controls the water supply and sewers for all of southeastern Michigan to Coleman Young's good reputation in Lansing from his Senate days. Several people also suggested that the state's commitment hinged on the strength of the city delegation in the legislature.

Elite perceptions of relatively good state–city relations in both Georgia and Michigan during the early years of the black mayors are borne out by the record. From 1973 to 1977 there was not one single legislative enactment or court decision in either state that could be construed in any way as inimical to the interests of Atlanta or Detroit. No special acts or discriminatory rulings sought in any way to constrain the mayors of the two cities or their new majorities.[10] Both states in fact enhanced the fiscal capabilities of their local governments during this period, Georgia by authorizing local option sales and income taxes in 1975, and Michigan by increasing the amount of money available for state shared revenues for local governments in 1976.

An inspection of trends in state fiscal aid to Detroit and Atlanta as a percentage of these cities' total revenues shows that in both cases the state contribution actually increased slightly during the black mayors' first terms (see Table 5.2). In neither case was state aid a function of federal intergovernmental transfers. The contribution made by the federal government to the two cities fluctuated both in percentage and absolute terms during the period between the 1969 and 1975 fiscal years, whereas the growth of state aid, except for a single year during Massell's mayoralty, was relatively steady.

With the exception to some extent of the activists in the Atlanta metropolitanization movement, white elites did not view state action as a way of reasserting or protecting white interests in the central cities. Certainly neither state legislature took it upon itself to strip its black-run cities of power. Indeed, such proclivities, if they exist at all, may never find an outlet in these two states. Both legislatures are among the leaders in the nation in number and percentage of black legislators. Georgia led the nation in 1976 with 22 blacks in its Assembly and Senate and was third in percentage of blacks in the legislature with 9%. Michigan ranked fourth with 15 black legislators and second in percentage terms with 10% (Joint Center for Political Studies, 1976). These small but increasingly

---

[10]Information is based on summaries presented by the International City Management Association for the years 1973 through 1977.

TABLE 5.2
Intergovernmental Aid in Atlanta and Detroit, 1969–1976
(in Millions of Dollars)

			Atlanta							Detroit				
Fiscal year	Mayor	Total local revenue	State aid		Federal aid		Mayor	Total local revenue	State aid		Federal aid			
			Amount	As % total	Amount	As % total			Amount	As % total	Amount	As % total		
1969–1970	Allen	$ 98.4	$ 4.7	5	$ 4.1	4	Cavanagh	$410.9	$ 54.2	13	$ 44.4	11		
1970–1971	Massell	107.9	4.1	4	5.6	5	Gribbs	500.7	63.1	13	96.1	19		
1971–1972	Massell	152.2	9.4	6	10.4	7	Gribbs	581.4	73.6	13	132.1	23		
1972–1973	Massell	156.3	9.5	6	15.8	10	Gribbs	665.3	81.6	12	186.8	28		
1973–1974	Massell	204.5	8.4	4	45.6	22	Gribbs	674.1	99.2	15	138.4	21		
1974–1975	Jackson	204.6	14.9	7	38.5	19	Young	661.0	100.0	15	139.8	21		
1975–1976	Jackson	203.9	23.3	11	22.7	11	Young	781.9	118.2	15	214.0	27		

Source: City Government Finances (1969–1976).

significant racial cohorts form a bulwark against the potential mistreatment of black-governed cities and make an appeal to the state by white elites for relief a strategy fraught with problems.

SUMMARY: THE ALLOCATION OF WHITE-CONTROLLED RESOURCES

An examination of various strategies of response to black rule at the elite level in the white communities of Atlanta and Detroit suggests that important resources controlled by whites, which would normally have been available under white rule to the city and city government, were also available under black rule. Although a number of politically prominent whites withdrew from the local political arena, strategies of adjustment in the law, business, and financial sectors were predominantly cooperative or maintaining in character. Few people or firms thought of abandoning the city or of mounting serious efforts to challenge the new mayors electorally or to obstruct their administrations.

This is not to say that the resources made directly available to the mayors or the investments in the economy of the two cities provided an embarrassment of riches from the private sector, that the availability of white resources made the difference between effective and ineffective black government, or that such resources necessarily guaranteed a level of economic and social health that black-controlled resources alone could not have sustained. But the adequacy and the impact of the white resources expended in the two cities, even if measurable, are not at issue. *What is important is that the flow of resources from the white elite community for both public and private purposes that had existed under white rule did not seem significantly altered by the transition experience.*

Let me summarize some of the particular sorts of resources white elites made available to the mayors as well as those investments that had putative impacts on the health of the cities the black mayors governed.

1. *Money.* The funding of both mayors' first campaigns came mainly from prominent whites in the business sector. Fund-raising was directed by white businessmen and lawyers. In Detroit a committee of white business figures and others also raised and managed a separate fund for use by the mayor to furnish the official mansion, to entertain for civic purposes, and to provide money for travel expenses of people brought to the city to interview as prospective mayoral appointees. Money was also supplied by major corporations to New Detroit to implement its minority-focused programs, and Detroit banks were active in helping to finance the city's projected deficit.

2. *Jobs and taxes.* In both cities business supported downtown development through the promotional efforts of Central Atlanta Progress, Detroit Renaissance, and the Chambers of Commerce. Numerous instances of plant renovation, decisions not to move, new buildings undertaken, and business expansion occurred, suggesting that investment, at least among the leading firms, continued during black rule.

3. *Prestige.* Major business elites in Atlanta and Detroit accompanied their mayors on business-seeking trips out of town. In Detroit they sat on the board of New Detroit, and a number were active in vocally supporting the mayor's effort to see the city through its fiscal crisis. Atlanta elites lent vocal support to Maynard Jackson during the sanitation strike.

4. *Expertise.* Whites were available to play advisory roles to the black mayors and to fill a variety of governmental and public service positions. In both cities the mayors were able to find apparently qualified whites to head administrative agencies and to serve on task forces, boards, and commissions.

In short, the pattern of white elite responses to transition revealed no systematic withdrawal of resources, and black rule did not demoralize the politically displaced group or drive its members, with their wealth and experience, from the city.

6

Metropolitan Reform and the Limits of a Subversive Strategy

Ever since the development of transportation technologies made long-distance commuting a daily possibility—thus putting into motion a great population movement out of central cities in search of a suburban idyll—the desire to bring the entire metropolitan area under the jurisdiction of a single local government has occupied a central place in the dreams of urban reformers. Their particular schemes have ranged from the imposition of full-scale metropolitan governments to the piecemeal strategies of annexation and single-purpose functional or planning districts.

Motivations for effecting such metropolitan arrangements have been varied. Reformers have viewed metropolitan government as a device for expanding the central city tax base, for drawing middle-class civic and political talent from the villages along the suburban fringe back to the city, for rationalizing the planning and delivery of public services and realizing economies of scale, and finally, as we saw in the case of nineteenth-century Boston, for diluting the power of those groups that remain in the central city, bound in place by poverty or color but growing in local political strength by dint of numbers. In recent years the movement for metropolitanization has gained impetus from the federal government itself, principally through its requirement that certain projects needing federal funds be reviewed by an areawide planning body to ascertain their metropolitan implications. The mix of motives—the search for tax revenues and equity, the quest for economy and rationality, and the desire to subvert the power of disadvantaged groups—makes the analysis of the issue a complex undertaking. Metropolitan reform efforts in Atlanta and Detroit are no exception.

METROPOLITAN REFORM AS A RACIAL ISSUE

The implications of metropolitan arrangements for growing central city black populations have long been a concern among blacks. In the early 1950s metropolitan reorganization efforts in Cleveland, St. Louis, and Nashville, all unsuccessful, met with strong black voter resistance. In the 1952 effort to annex portions of Davidson County to Nashville, the curbing of growing black political power in the central city surfaced as an explicit argument by white proponents of the plan. In Virginia in 1956, blacks rejected a proposal supported by a majority of white voters to merge the city of Newport News with surrounding suburban areas. Central city voters of both races rejected similar plans in Nashville once again in 1962, and in Tampa in 1967. In 1969 black leaders fought the city–county merger of Indianapolis and Marion County, a reform effected entirely by the state legislature without a local referendum. The only recent cases in which a majority of black voters supported metropolitan reforms occurred in Jacksonville, Florida (1967) and Lexington, Kentucky (1972), where blacks were promised increased representation in local government through a change from at-large to ward-based aldermanic elections.[1] However, black opinion has generally held, as Gary Mayor Richard Hatcher put it in 1970, that attempts to develop metropolitan government seek "to mute the black votes" and enervate the inner city (Hawley, 1972, p. 5). Such a view finds at least circumstantial support in the fact that four out of the five most recent city–county mergers have occurred not in the west, the region where the general tradition of reform maintains its strongest grip, but in the South in cities with black populations approaching a critical political mass.

Academic observers have also been accustomed to interpreting conflicts over metropolitan reform in racial terms, even though white majorities as well as black in both central cities and suburbs have rejected

---

[1]For a summary of these various reform efforts see the study by the Advisory Commission on Intergovernmental Relations (1974, pp. 101–103).

more of such proposals than they have accepted. In 1958, in the very infancy of black urban political power, Morton Grodzins warned that the annexation of white suburbs would be one countermeasure to black political domination: "This will be a historic reversal of the traditional suburban antipathy to annexation [p. 14]." Others have not been convinced that local white voters, particularly in the suburbs, would support merger with black-dominated central cities but argued nevertheless that federal or state efforts to impose metropolitan decision-making structures on cities and suburbs would have the same effect of frustrating black political power (see, e.g., Piven and Cloward, 1967; Friesema, 1969).

Attempts to reconcile black political interests with metropolitan reform have not been entirely lacking. We noted, for example, that proponents of the successful city–county consolidations in Columbus (Georgia), Jacksonville, Indianapolis, and Lexington sought to build in guarantees of minority representation that blacks did not enjoy under the old structure of government. Thus, although black strength as a citywide voting bloc was diminished by metropolitan reform in each of these four cities, black representation on the new city councils increased after reform in all but Jacksonville (ACIR, 1974, p. 103). Academic reform proponents have also begun to address the problem of black interests. Willis Hawley (1972), for example, has argued that the increased capacities for planning and revenue-raising brought about by metropolitanization would enhance the capacity of local government to respond to the very issues—housing, employment, schools—in which blacks have a major stake.

In any metropolitan reform, however, blacks in the central city face a trade-off. The legitimate benefits accruing to the central city (and to the suburbs as well) in terms of tax and revenue equity and service rationality must come at the expense of black chances to control the top office in central city government. In cities like Detroit and Atlanta, which already have black mayors, the issue of metropolitan reform assumes particularly troublesome dimensions. Recognizing the threatening nature of metropolitanism for black mayors so recently installed, some proponents of reform in those cities tendered plans festooned with the language of rationality and efficiency, anguishing over the potentially subversive implications of their ideas for the future of black power. Others cloaked their essentially subversive intent in the publicly acceptable terms of fiscal necessity. Some assigned priority to reform or to black political interests and, fully aware of the conflicts between the two, simply planned to bear the costs of their preference. Virtually no one, however, was primarily and openly interested in metropolitanization as a subversive response to transition. Nevertheless, the adverse impact on black voting strength and morale of any metropolitan reorganization was clear and widely recognized.

METROPOLITAN DEVELOPMENTS IN ATLANTA AND DETROIT

The greater Atlanta and Detroit areas both look back on histories of modest ground-breaking initiatives in metropolitan reform. Atlanta's Plan of Improvement, implemented by the Georgia General Assembly in 1952, represented the first time in the United States that a locally appointed government commission developed and successfully secured a major metropolitan reform program. The Plan of Improvement called for liberalized annexation procedures favoring the central city, county-wide property appraisal, and the reallocation of services provided by Fulton County and the City of Atlanta in such a way as to avoid duplication (Holland, 1952). Annexations immediately resulting from the Plan of Improvement added 81 square miles to the city and reduced the black population from 35% to 30% (AC, 30 March 1975). In Detroit the Supervisors Inter-County Commission, established in the metropolitan area in 1954, was the first council of governments. Under the impetus of federal incentives, councils of governments—essentially voluntary metropolitan bodies that perform advisory planning functions—greatly increased in number nationally during the 1960s, copying the pioneer effort in Detroit.

Aside from these initiatives, however, the story of metropolitan reform in the two cities is a more typical one of failures and frustrations. As in other places, proponents have engaged in endless studies over the years. Many of their resultant proposals have made their way into the state legislature, only to languish or be defeated. Racial considerations as an acknowledged factor in such reforms seem to have surfaced only in comparatively recent times, adding complexity to the debates on the issue but by no means dominating them.

Atlantans have been thinking about metropolitan government in one form or another at least since 1912, when a New-York-based public administration research organization recommended the consolidation of Atlanta and Fulton County in the interests of economy and efficiency. A dozen years later the Georgia State Assembly passed enabling legislation permitting the merger of city and county governments. Between 1933, the first year in which legislation was introduced providing specifically for the consolidation of Atlanta and Fulton County, and 1966 there were three major annexation and merger bills debated, three popular referenda on the subject, and at least seven different studies produced by various citizens' commissions and consultants.

The racial issue first became important in 1966 when the legislature passed an act allowing the city to annex a well-to-do white suburb called Sandy Springs, pending voter approval there. To counter the proposed infusion of white voters, Senator Leroy Johnson, the first black in modern times to serve in Georgia's upper chamber, successfully amended the bill to include the mostly black area of Boulder Park in any annexation scheme. Referenda were duly held in both suburbs; annexation was approved in Boulder Park but rejected in Sandy Springs, thus killing the entire plan.

After a Georgia Supreme Court decision in 1968 struck down the annexation procedure developed in the 1952 Plan of Improvement, legislative activity to restructure the metropolis intensified. Between 1967 and 1975 an additional seven bills calling for city–county merger, annexation, or some other form of metropolitan reorganization were introduced. None made its way through the intricacies of Georgia state politics. Additional impetus for this flurry was an amendment to the state constitution in 1972 allowing counties to provide a greater range of services (even if they duplicated those that cities offered to county residents), thus effectively relegating the moribund Plan of Improvement to the legislative junkyard. The only important existing metropolitan body in the Atlanta area during the mid-1970s was the Atlanta Regional Commission, established in 1971 to perform advisory planning functions for the seven urbanized counties surrounding Georgia's capital.

The history of metropolitan developments in Detroit is a less active one, reflecting a greater consensus on the desirability of local efforts to achieve a limited governmental mastery over the region and greater confidence in the structures available to perform such functions. The formation in 1954 of the Supervisors Inter-County Commission (SICC)—the council of governments—actually followed the organization in 1947 of a metropolitan regional planning commission. The planning efforts of the latter body, advisory in nature, provided the subject for discussions among local officials on the SICC.

Since the activities of these two bodies were so closely intertwined, a civic committee devoted to metropolitan reform recommended in the early 1960s that the SICC and the regional planning commission merge. They did so in 1968, reconstituting themselves as the Southeast Michigan Council of Governments. SEMCOG, as this body is known, develops plans in the areas of transportation, land use, criminal justice, sewage, water resources, housing, and the environment for the seven-county region surrounding Detroit. Although it may not implement its plans, it serves as the A-95 review agency for the metropolitan area, forwarding its purely advisory recommendations to state and federal funding agencies.[2] Membership in SEMCOG is voluntary. Of the 240 local government units in the 7 counties, only 112 were members in 1977. The work of the council is supported through the provision of research funds and staff by a nonprofit organization called the Metropolitan Fund, whose resources and chairpersons have been drawn from major Detroit corporations and labor unions. Several additional single-purpose metropolitan bodies both plan for and administer certain regional services including mass transit, parks, and water. Racial factors did not play a major role in discussions of metropolitan government until the advent of busing to achieve school integration in the early 1970s and the subsequent election of the black mayor.

METROPOLITAN INITIATIVES UNDER THE BLACK MAYORS

During Maynard Jackson's first term nearly everybody who was anybody in Atlanta had his or her own plan for metropolitan reform, even the mayor himself. The Commerce Club, under the leadership of banker Augustus Sterne, was awaiting the results of a study conducted by a private consulting firm, Research Atlanta, on the consequences of a merger between the city and Fulton County. The mayor, searching for ways to expand Atlanta's tax base, acknowledged the need for limited annexation and supported a plan that would add to the city the sparsely populated but tax-rich Fulton Industrial District. At least two prominent county commissioners were promoting separate plans involving variations on a federal scheme of organization for Fulton County. The former governor of Georgia, Carl Sanders, who served in the state house from 1963 to 1967, was leading an effort to create a five-county metropolitan federation, a proposal that would require a state constitutional amendment. And in 1977 Governor George Busbee appointed a study commission—another in a long line—to explore the interrelatedness of the governments of Atlanta and Fulton County and to offer solutions regarding service delivery and financing. Refusing to endorse any particular plan, but supporting the principle of metropolitanization in varying degrees,

[2]A-95 refers to a circular issued by the federal Office of Management and Budget implementing a provision in the Demonstration Cities and Metropolitan Development Act of 1966 requiring that all federal grant applications made by local governments be reviewed by an areawide planning body before being sent to Washington.

the Chamber of Commerce, the League of Women Voters, and Central Atlanta Progress added their weight to these uncoordinated efforts by issuing occasional calls for reform. As one prominent reform activist (I-229) observed in an obvious understatement, "There's a lot of duplication of task forces and committees, a lot of the right hand doing things the left hand doesn't know about."

Several aspects of this frenetic activity are important. One was the nearly universal agreement among prominent central city whites on the need for some type of metropolitan solution to Atlanta's various problems of planning, financing, and coordination. Conversely, black elite opinion during this period was almost entirely negative regarding metropolitanization, with the exception of those few who favored Jackson's modest efforts to acquire the Fulton Industrial District for the city. In the suburbs white politicians and their constituents formed a solid phalanx in opposition to all proposals. Thus, cleavages on the issue not only followed racial lines but also divided central city whites from suburban whites. Furthermore, compounding the patterns of division, central city white advocates of reform found it impossible to coordinate their specific efforts. Nevertheless, it is important to note that every one of the plans being discussed by central city whites, save that which called for the annexation of the Fulton Industrial District, had the effect of reducing the black voting population in the reconstituted city to less than 50% (see Table 6.1). We shall examine these patterns of support and opposition in more detail shortly.

Metropolitan initiatives in Detroit during Coleman Young's first term were much more limited than in Atlanta, a function to some degree of the existence of a significant number of regional single-purpose au-

TABLE 6.1
The Impact of Proposed Metropolitan Reforms on Black Voting Strength

	Resulting voter population	
Boundary change	Black	White
Annex Sandy Springs (S.S.) to Atlanta	45.7%	54.3%
Annex Fulton Indus. Dist. (F.I.D.) to Atlanta	51.7	48.3
Annex S.S. and F.I.D. to Atlanta	45.6	54.4
Annex unincorporated N. Fulton to Atlanta	44.9	55.1
Annex unincorporated S. Fulton to Atlanta	48.9	51.1
Consolidate Atlanta and uninc. Fulton	42.7	57.3
Consolidate Atlanta and all of Fulton	37.9	62.1
Establish five-county metro federation	21.0	79.0

Source: League of Women Voters of Atlanta–Fulton County (1975, p. 9).

thorities already administering certain services on a metropolitan basis. The principal effort was being led by William Ryan, a white state legislator from Detroit with a background in the labor movement, who represented a racially mixed constituency. Ryan's bill, H.B. 5527, grew out of a proposal by the Metropolitan Fund. Introduced in July 1975, it sought in effect to arm SEMCOG with governing authority. Ryan hoped not only to make membership by local government in the newly constituted body mandatory, but also to vest in it decisive planning powers over the single-purpose agencies as well as several other functions. The bill failed to emerge from the House Urban Affairs Committee and died with the end of the 1976 legislative session. Ryan reintroduced the bill in 1977 without success.

Patterns of support and opposition, both to the specific bill and to the idea of metropolitan reform, were similar to those in Atlanta. Led by the Metropolitan Fund, Detroit's white labor and corporate elite united to support the Ryan bill. Most prominent blacks, including the mayor, were vehemently opposed both to the bill and the general concept. They were joined in opposition by white suburbanites, who feared the integration of their schools. SEMCOG itself also opposed the Ryan bill, arguing that regional voluntarism was adequate to deal with the problems of the metropolis. Significantly, the chairman of SEMCOG during this period was a black man, the head of the Wayne County board of commissioners.

PATTERNS OF SUPPORT, OPPOSITION, AND MOTIVATION

No one interviewed in Detroit or Atlanta had the least difficulty in outlining the broad patterns of support and opposition on the complex issue of metropolitan reform, even though in Atlanta, in particular, reform connoted so many different sorts of plans. Furthermore, there was almost complete consensus regarding the nature and implications of these patterns.

In Atlanta it was universally agreed among those interviewed that suburban white opposition to metropolitanization would be virtually impossible to breach, making it necessary for any reform to be imposed by the state legislature, without provision for popular referendum. The prospect of school integration through busing was seen to be the key to suburban sentiment. Indeed, some 40,000 white pupils had left the Atlanta city school system for suburban schools in the decade prior to Maynard Jackson's election. By 1974 the school system in the city was 84% black. Detroit elites also saw suburban fears of school integration as a major barrier to regional government, although black opposition loomed as the more important factor in their calculations of its short-run chances.[3]

In a few cases black opposition tempered the enthusiasm of white elites for metropolitan reform or even caused them to oppose it. A spokesman for Atlanta's business community (I-216) suggested that "victory wouldn't be worth it if blacks didn't accept it." And a major white politician in Detroit (I-112) refused to support any proposal for metropolitanization "as long as black leaders oppose it." Representative Ryan himself sought to allay both suburban white and central city black fears by arguing that his proposal was a minimal one designed only to render certain regional services while preserving the right, as he noted in his interview, of "Dearborn's white power and Detroit's coalition power" to exist.

To Detroit elites, generally, black opposition seemed intractable and, on the whole, understandable. Thus, to a large extent black opinion represented the proverbial immovable object before which metropolitan reform ambitions would necessarily have to give way. In Atlanta, however, a number of white elites were convinced that blacks could be "sold" on the idea once the rationality of reform was made clear. "My businessman's background," one said (I-212), "leads me to say that I can't conceive that black leadership wants the community to go to hell financially. They have to realize that some kind of trade-off is necessary. They should be interested in getting some balance between representation and financial viability." A leader in the reform movement (I-229) made a similar point: "I think it's going to have to be sold to blacks by telling them they're going to be locked into a core city that won't be able to afford the services they want. A lot of blacks are going to support this." Maynard Jackson's willingness to seek the annexation of the Fulton Industrial District led other whites to believe that black leadership could eventually be moved on the issue (I-25).

Paradoxically, the belief that Atlanta's black leaders might come to accept the rationality of reform coexisted with the relatively widespread assumption that many advocates of metropolitanization were moved by racial motivations. Elites who were interviewed, both those active in reform efforts as well as observers, were careful to deny that racial considerations played any part in their own feelings about the issue.

---

[3]Although black opponents of metropolitan reform were extremely vocal, white suburban opposition seemed more organized. Alerted to the nature of the Ryan bill by a letter from Oakland County Commissioner Dana Wilson ("a 'backdoor' attempt to *bus your children*," the letter said), white suburbanites organized a recall drive against the bill's cosponsor, Philip Mastin, a representative from Hazel Park. The effort failed, but it served to place state legislators on notice.

Instead, they referred to an unspecified "they"—by whom they meant advocates of reform other than themselves—who sought to exploit metropolitanization for the purpose of regaining white power. A white county commissioner (I-230), for example, suggested that "they feel blacks haven't been the best for Atlanta government and they'd love to figure a way out to get power back for whites." A banker (I-219) averred that "annexation is surely a rallying point for some to take power away from blacks," and a white on the city council (I-224) interpreted "all this sudden concerted attempt to get [reform] as a reaction to the new racial balance." Another respondent argued that race was an important stimulus to recent reform efforts but sought to minimize the role of racial considerations historically by pointing out that the issue had hardly arisen prior to 1970 (I-25).

Atlanta elites pictured their own motives for supporting metropolitanization as strictly rational. That is, they explained their interest as an effort to expand the city's tax base and to eliminate service duplication and fragmentation. They were on the whole well informed, referring often to Toronto's metropolitan federalism or Minneapolis–St. Paul's metropolitan tax-sharing as models to emulate. A few argued too that annexation would bring "the better-educated suburban population" back into the decision-making processes of the city, a hope, we may recall, that fueled Brahmin interest in the early movement for such reforms in Boston (I-29, 210).

For Detroit elites the conflict over metropolitan reform had an ironic quality: Although blacks appeared intransigent in their opposition, no one in Detroit suggested, as people in Atlanta had, that the desire to dilute black power lay behind the call for reform. The move for metropolitanization was clearly perceived as a liberal cause aimed only at tax base expansion and service rationalization. "Metropolitan consolidation is not aimed at racial subversion," said one politician (I-112), "especially the Ryan bill. . . . The liberals in favor of it are really worried about what resources the city run by blacks will have."

ASSESSMENTS OF THE FUTURE
OF METROPOLITAN REFORM

No trait has emerged so clearly in the exploration of elite adjustments to racial transition as the fundamental realism of white expectations. This was the case, for example, regarding estimates of the duration of black dominance in the two cities, and it was the case as well in assessments of the prospects for metropolitan reform. It is true that some of those interviewed maintained strong, if improbable, hopes or indulged in the construction of unlikely scenarios—for example, that black leaders could be "sold" on metropolitanization (I-233)—but none of these achieved the stature of fantasies in which the white community would somehow be rescued from its fate. Rather, these modest scenarios of how to win metropolitan reform or of who might challenge the incumbent mayor in the next election might best be understood as preliminary speculations, unconstrained by responsibility to the laws of political probability. When, however, these observers actually came to the calculation of odds, they did so virtually without illusion. Thus, it was the almost unanimous opinion of elites in both cities, despite their generally strong support of the concept, that metropolitan reform of any sort stood little chance of passage or adoption in the near future. Speaking of the activists in the reform movement, one Atlanta politician (I-234) commented, "They've got all their plans and task forces. They're trying to figure out how to get the country boys [in the legislature] to go along. But it ain't going to happen." Detroit elites offered similar prognoses (I-14, 112, 123, 134).

Several factors seemed to govern these assessments. One involved the apparent need to obtain approval from the U.S. Attorney General, as required by the Voting Rights Act of 1965, for any reorganization plan that enlarged or reduced the voting population in any of the affected political subdivisions such that a racial minority was potentially disadvantaged politically.[4] Any proposal with a racially discriminatory purpose or effect is barred by the act. Although advocates of metropolitan reform in Atlanta understood that an equitable reorganization plan could be designed, they were not sure that any plan under discussion could meet the burden of proof of nondiscriminatory impact, particularly given the political reality of black dominance in the central city.

A closely related factor was the sense that reform could not be imposed on an unwilling black community. "Annexation requires neutrality," noted a Detroit politician (I-17), "and that isn't going to happen." An Atlanta real estate developer (I-213) saw the costs in terms of racial amity as simply too high: "In a realistic way it's not possible to expand the city. The only way would be if there were to be a real racial row. . . . That's no way to get expansion, even though I want it very much."

Pessimism regarding the future of reform was also encouraged by the perceived lack of interest in the issue in the respective state legislatures. Neither assembly felt any evident responsibility to "save" their major cities from black rule through metropolitan devices. Out-state

[4]P.L. 89–110, Sections 3 (c) and 5.

apathy, suburban and black opposition, and what amounted to guber-
natorial neutrality or very modest support produced an additional legislative immunity even to the most carefully constructed rationality arguments
of the liberal reform advocates.

Elites in both cities finally regarded the passage of reform as unlikely because of the history of incremental establishment of single-purpose
metropolitan authorities. With transportation and advisory planning already metropolitan functions in both places (though the transportation
system is limited in Atlanta to Fulton and DeKalb counties), and the
metropolitan administration of water, sewer, and some recreation and
health services in Detroit, the extreme urgency of regional government
is unclear to all but its staunchest supporters. As one Detroit business
executive (I-118) summed up, "Regionalization you can get away with,
but metropolitan government—well, that puts everyone in a tizzy."

CONCLUSIONS: METROPOLITAN REFORM AND
POLITICAL ACCOMMODATION

There can be little doubt about the nature of the consequences of
successful metropolitan reform for the short-run prospects for black
political power in Atlanta and Detroit. With racial voting blocs so nearly
equal in size and with voting behavior closely tied to race, the addition
to the city through annexation or merger of only a few thousand white
voters could easily restore the pattern of power to its pre-1973 dimensions. Black control would undoubtedly be challenged in an enlarged
city, and the chances for success of such a challenge are probably high.
Alternatively, in a metropolitan federal scheme, in which city boundaries are not changed but service functions are reassigned to the new
metropolitan umbrella jurisdiction, the black governors of the central
city would stand to lose control over key municipal agencies, such as
those responsible for policing, renewal, and housing. At least in the
short run, most types of metropolitan reform offer few apparent gains
for politically dominant blacks that might offset potentially severe losses.

If the subversive impact of reform for blacks is clear, however, the
subversive intent of whites interested in metropolitanization is less so.
Although the desire to undercut black power was widely assumed to be
a central motivation of many of Atlanta's reformers, the advocates of
reform could point to a host of other reasons that justify metropolitanization, each with a substantial and thoroughly respectable pedigree. To some
extent the very mix of rational reform and venal motivations had an
enervating effect on the efforts of those who fit the classic mold of the

civic reformer. Beset by concerns about the "fairness" of reform in the prevailing political circumstances, the advocates of metropolitanization as a rational reordering of the urban region tempered their efforts. Perhaps more than any other factor, such concerns explain the lack of mobilization and coordination of the reformers in both cities.

The sense that reform would not be entirely fair to blacks who have just won political power—or at least that reform should not be imposed on an unwilling black community—suggests that transition cannot be widely regarded as so great a threat as to cause a united white community to throw all considerations of social costs to the winds in seeking to regain power. Neither is transition seen to be so fearsome as to galvanize the white community for united action. White urban Americans, at least at the elite level in Detroit and Atlanta, could scarcely be said to be engaged in a race war. The fact that metropolitan reform did not enjoy sudden success in those places—a function both of the failures and doubts of central city whites as well as the opposition of white suburbanites and blacks—indicates a search for accommodation rather than a compelling quest for racial dominance.

Central city white elites signalled, by their lack of decisiveness and cohesion, a basic willingness to live under black rule. Although their attitudes regarding reform were generally favorable, the issue did not command such priority that they were willing to ignore black opposition. Racial conflict over metropolitan reorganization will not necessarily diminish. But at least during the first terms of the black mayors, when the novelty, if not the shock, of transition was most acute, white leaders were unwilling to seek to push the issue to a showdown confrontation between the races. The issue of metropolitan reform thus illustrates as clearly as any the limits in Detroit and Atlanta of a subversive strategy.

7

Explaining the White Response to Transition: The Impact of Black Rule

In the autumn of 1977, Coleman Young and Maynard Jackson easily surmounted electoral challenges to win second terms in their respective city halls. The mood in both cities seemed optimistic. Economic recovery from the recession of the early part of the decade was well under way, as residential and downtown construction surged. Political observers spoke of the emergence in both places of an "unusual coalition" of blacks and big business (*NYT*, 9 Sept. 1977, 6 Oct. 1977). To all appearances the apprehensions of white elites over black rule had evaporated in a flush of good will (I-237).

The swiftness with which white fears had been largely allayed and transformed to acceptance—even in some cases satisfaction—seemed a drastic telescoping of a process that had taken far longer in turn-of-the-century Boston. Black rule in the modern cities appeared almost unremarkable a scant 4 years after the initial black victories. Maintaining and cooperating strategies were the predominant responses of white economic and political elites, whereas subversive efforts and electoral contestation were disorganized, moribund, or entirely lacking. Except for the decisions of a few white mayoral hopefuls not to run for city hall, there was little evidence of elite or corporate withdrawal from the city or its affairs. The adjustment of white elites to the new configuration of power was at all times peaceful and notably free of rancor. Although white elite opposition to the black mayors was not widespread, that which did develop was predicated for the most part on a search for accommodation rather than conquest. By all indications, then, transition to black political dominance had not only proceeded smoothly, but in such a way as to preserve for the new governors a high degree of access to the various resources controlled by white elites. It is surely doubtful that any observer in the late 1960s, schooled in the well-established lessons of American race relations, would have predicted accurately the essentially yielding, acquiescent character of the white elite response to transition.

In an important sense it seems evident that political transition in the two cities has taken place in and further reinforced a culture of accommodation, which may be seen in the broad attitudinal changes that emerged in the aftermath of a long history of intense racial conflict. The culture in each city has certain specifically local origins and characteristics, but is nevertheless the clearly identifiable offspring of more general American traits. It is notable for the relative absence of bitterness on the part of the displaced and the lack of vindictiveness among the leaders of the new political majority. In its local versions in Detroit and Atlanta the culture of accommodation is a product of white elite judgments and adjustments as well as of the sensitivity of the new victors.

The culture emerges from the recognition that blacks and whites living in the same city share some very basic interests. Its major dimensions consist of normative guidelines for governing relations between the races in a particular stage of development. The generation of norms of accommodation and their application to ethnic political conflict have not been uncommon in American politics, but the extension of the culture to black–white relations is new.

The accuracy of a description of American ethnic political relations as governed eventually by emergent norms of accommodation is, of course, dependent upon a developmental perspective. Initial stages in interethnic relations in America—often of substantial duration—have historically been marked by unfettered brutality, repression, and discrimination against subordinate groups by the dominant ones. The decimation of the Indians, the treatment of blacks, Catholics, Orientals, and Hispanic Americans, and the more subtle humiliations imposed on American Jews all provide ample illustrations of American conformity to a virtually worldwide attraction to variations of a politics of ethnic repression and discrimination. Yet in all of these cases the relationship between the subordinate group and the dominant society has undergone transformations leading at least to peaceful modes of contact and at best to more equitably structured interactions. And what is striking moreover about these transformations is that they have not occurred—except in the case of the American Indians—as a consequence of conquest by one group over the other.

Cause and effect here are by no means simple to sort out. In part, groups seem to work out patterns of accommodation from the sheer exhaustion of maintaining hostilities. In addition, nationalistic pressures in wartime, the broader range of opportunities generated by an expanding economy, and growth of the suburban escape valve may also have contributed historically to diminishing interethnic animosities. In times of lessened interethnic tension, subordinate groups may find fewer barriers to successful political mobilization and thereby work their way into the political system.

On the other hand, however, the key event in the diminution of dominant group hostility toward a subordinate ethnic group has often seemed to be the successful capture of important political office by the latter. In such a situation of transition, members of the formerly dominant group must, if they do not migrate, acknowledge a certain degree of formal interdependence with the newly powerful group, for the fortunes of both groups depend substantially on the quality and character of the governing enterprise. This is likely to foster a sense of mutuality in the political community. As Robin Williams, Jr. (1977) has written, "When individuals of different ethnic groupings are placed in relationships of positive interdependence in which successful performance requires cooperation, there will be an increase in helping, in mutual liking [p. 277]." It would appear then that transition marks a critical stage in ethnic political relationships by helping to stimulate the emergence of a culture of accommodation.

The norms of the culture of accommodation do not by any means suggest an end to ethnic-based political conflict nor (even more unlikely) a societal willingness to apportion the fruits of politics more equitably to take into account the newly emergent group's entry into the club of electoral competitiors. *Accommodation* means rather that the subordinate group that has competed successfully in electoral politics acquires a certain political legitimacy and measure of respect it did not have before. It must now be taken seriously, for it has demonstrated a capacity to take for itself the offices that were heretofore the preserve of the dominant group.

Judging on the basis of both the interview material from Atlanta and Detroit and the actual behavior of elites in those cities, it would appear that certain normative themes are characteristic of the culture. Physical and verbal violence for political purposes is essentially nonexistent. Weapons important in the arsenal of white politics in those places only a few years ago, such as race-baiting, mob action, police brutality, and incendiary speeches, have vanished from the scene. Optimism rather than despair about the future of the city emerges as a rhetorical refrain.

Old patterns of resistance (George Wallace standing in the school-house door), which characterized the response of the dominant group under challenge, have given way to bargaining strategies once the group is out of political power. Cooperative interracial relationships for mutual benefit—described as coalitions of the private and public sectors or of business and "neighborhoods"—have become ruling models. Elites of both races are more prone to speak of the convergence of racial interests and the search for means to accommodate this new mutuality than of autonomous racial development or racial conflict. Compromise rather than conquest or intransigence is the norm for dealing with racial disputes, and to encourage such accommodation both cities have witnessed the proliferation of institutionalized and quasi-formal settings for carrying out "dialogues" between the leaders of the two races.

Above all, the culture of accommodation legitimizes the new arrangement of political power by explaining it as a product of democratic processes. The sanctity of electoral outcomes, majority rule, and "fair play" are all norms readily employed by white elites in seeking to come to terms with black government.

Neither this rhetoric of democracy nor the other elements of accommodation are to be taken lightly; the culture is neither posture nor mere verbal facade. It describes a set of rules for the transfer of formal power and adjustment thereafter by which these white and black elites live and beyond whose limits no one seriously thinks to go. It is true that this culture appears to come into play in American society primarily once a particular stage in the development of ethnic political relations has been reached, namely when competition is at last pursued in the electoral arena. This suggests that a major reason for obeisance to the democratic and accommodationist norms on the part of a displaced group is the need to create or reinforce a climate of tolerance that will guarantee their survival in a system in which they are no longer absolutely politically dominant. But whatever the motives, these cultural formulae seem to work to diminish the intensity, volatility, and unpredictability of ethnic political competition.

EXPLAINING TRANSITION

To suggest that acquiescence to transition on the part of displaced elites is a direct product of a benign culture of accommodation is only to begin to explain why the white adjustment to black rule has taken the form it has. The crucial question to address is how the culture of accommodation is sustained in this situation. What in short can explain why white elites have accepted so easily the logic of democracy—namely political displacement—and a cooperative role in the new order? If we dismiss the notion that pure virtue has shaped the dominant white responses, then we are left with three possible general explanations.

The first is that transition to black rule is essentially cost-free, if not even irrelevant, to white elites. To lose control of city government is simply to lose possession of a hollow prize. This explanation has several components. It assumes first of all that control of city hall has limited symbolic value for whites. More important, perhaps, it assumes that city government cannot hurt the interests or well-being of white elites in any significant way. City government is so inherently weak that it cannot regulate important white activities or redistribute public goods and resources to the black community. Finally, it assumes that city government is essentially incapable of advancing white interests, whatever they might be; therefore, to lose control of government is to lose no advantage. In short, this explanation suggests that whites can afford to sustain the benign norms of the culture of accommodation because control of city government does not materially affect their interests. However, although this is an important and, indeed, seductive explanation, careful scrutiny of the evidence suggests, as we shall see later in this chapter, that it cannot be sustained.

A second explanation, for which there is a firmer basis, focuses on a variety of environmental factors that both encourage acquiescence to transition and militate against resistance. These factors include national and local cultural forces shaping the character of race relations, the configuration of economic interests, and the nature of the local political and social structures. All of these appear to have converged in such a way as to create a climate conducive to the maintenance of a culture of accommodation. According to this perspective, which I examine in Chapter 8, white acceptance of transition is partly a product of liberalizing trends in racial attitudes. In addition, positive incentives in the form of economic benefit and social stability encourage acquiescence, and certain features of the political and social structures in the two cities serve to dissipate potential opposition.

A third possible explanation for the benign transition process, discussed in Chapter 9, begins with the assumption that politics in America is limited as an instrument of social change; therefore, certain very basic established interests are not likely to be affected greatly by any sort of political developments. This explanation differs from the notion that transition is cost-free because city government is powerless. It depends rather on an analysis of the role of politics generally in American life. From this point of view the debate over the degree to which city hall possesses inherent power is immaterial. American government in any jurisdiction and of whatever endowments observes certain restraints that guarantee at certain levels the preservation and integrity of groups that lose conventional electoral contests. Because of this, white elites can feel secure. Political defeat is not life-threatening, nor does it result in dispossession, exile, discriminatory or confiscatory taxation, or the withdrawal of civil liberties. Neither is the maintenance of privileged class status profoundly linked to control of political office, for the notion of expropriation is anathema to American political practice. In short, survival —and the survival of privilege—does not hinge on political victory. Losers in American politics can therefore live with their loss and survive in the new order virtually as well as they lived in the old. This does not mean that political office is unimportant at some level, but that the scope of American politics must be looked at in its proper perspective.

These three explanations for acquiescent white adjustment to black rule—that city government is irrelevant to white interests, that a variety of environmental factors encouraged acceptance, and that politics in America operates within a firmly established set of reassuring limits— are pitched at different levels of generality, from the specific elements of the environmental explanation to the more general considerations concerning the nature of power inherent in city government to the broadest analysis of the character of American politics. As we shall see, they are not equally valid. Yet with modifications each may contribute to understanding the transition process in Atlanta and Detroit; furthermore, they can help us to understand both historical and future ethnoracial political transitions in America.

DOES IT MATTER TO WHITE ELITES WHO RUNS CITY HALL?

The key to this plausible, yet inaccurate explanation for the easy nature of transition to black rule lies in an analysis of American urban government as an essentially powerless entity: City hall can neither profoundly hurt nor help, deprive nor enrich. It is based on the premise that the important decisions affecting life in the city are made by superior governments and by private sector firms. Thus control of the apparatus of local government is essentially irrelevant.[1] Since blacks seemed to have invested the mayor's seat with such substantial symbolic import, whites could afford to acquiesce to black victory in the interests of social peace without jeopardizing any important interests. The converse side of this argument is that if whites lost nothing of significance in losing city hall, then blacks gained nothing beyond the honorific gratifications of the mayoralty.

[1]A representative statement of this argument appears in Walton (1972, p. 201).

Before examining this explanation it is important to concede at the outset that city government in America is inherently weak. One-third of its revenues (and nearly one-half for large cities) come from superior governments, which have preempted the best tax sources. The local property tax, the major source of local revenues, is the least productive tax source and the least responsive to economic growth. Furthermore, it places big cities with their massive revenue needs in a weak competitive situation vis-à-vis suburban towns in the effort to attract major industrial and commercial tax producers.

In addition, governmental authority at the local level is so highly fragmented that mayors do not always control functions as crucial as public education, housing, welfare, and public health. Cities are creatures of their states, which determine, often in the most minute detail, the powers they may exercise. City governments can do very little on their own of a systematic nature to create large numbers of jobs, redistribute income, or provide great amounts of good housing.

All of these weaknesses notwithstanding, local government is still the major service provider in the public sector. The range and quality of services a city chooses to offer materially affect the quality of life in a city. City government not only shapes the nature of public services consumed but also influences the private sector. City hall increasingly has tools at its disposal to bring leverage to bear on private sector firms, influencing their hiring practices, their investments, and their location decisions. If the formal powers inherent in the mayor's office are modest, the vantage point the office provides for what might broadly be called "leadership" is not without consequence for special constituencies. Indeed, certain relatively novel strategies—affirmative action, preferential city purchasing—have begun to open up that allow mayors to exercise some modest, though undetermined, impact on income distribution in the cities. In short, if city government is weak in some respects, control of it is nevertheless increasingly a meaningful prize to be sought for certain purposes. In order to establish the nature of those purposes, it is necessary to determine both what whites have lost by losing majority status and control of government and what blacks have gained.

THE INFLUENCE OF BLACK GOVERNMENT ON SUBSTANTIVE BLACK AND WHITE INTERESTS

As the analysis in Chapter 4 suggests, white elites anticipated the transition to black rule in 1973 with fear. They believed that black rule marked the end of an era of more comfortable, understandable dimensions, and that they and their interests no longer resided at the center of political power in the city nor did their problems seem to shape the agenda of concerns. Atlanta businessmen in particular worried about the "image" of a city headed by a black. In both cities the change was understood not in terms of a turnover in the personnel of city hall but as a loss by one race to the other.

By 1978, however, the symbolic implications for white elites of a black in city hall had evidently receded in importance; patterns of acceptance had firmly taken hold. In contrast, interviews conducted in 1978 indicated that for blacks the importance of having a black in power to serve as a role model for members of the local black community and to provide evidence to the nation—indeed to the world—that a black could move among and even speak back to influential whites had not diminished. And in a variety of substantive areas the moral authority of a black leader was considered crucial to black gains.

If the blackness of the mayor no longer greatly mattered to white elites a few years after transition as a symbol of the change in their world, how much did it make a substantive difference to white and black interests? Some of these interests are such that black gains do not imply white losses; indeed, some interests apparently converge. Three stand out most sharply in these two categories: police behavior, relations with the federal government, and economic development. Other interests are in opposition, however, and must be viewed in terms of a modified zero-sum equation. Areas in which black gains imply, to some degree, white losses (the zero-sum equation is not a perfect one) include affirmative action in government, the distribution of local government contracts and business, and the autonomy of private sector hiring practices.

POLICE BEHAVIOR

When officials in the administrations of Coleman Young and Maynard Jackson were asked what difference, if any, it had made to black people to have a black mayor, they invariably mentioned improvements in police conduct as the single accomplishment most visible to the ordinary citizen. Several blacks in positions of major responsibility were able to recount stories of their own humiliating treatment by the police in the past. "But there's no police brutality toward blacks these days," said one high Atlanta official (I-240). "They treat people like people now." And a Detroit official (I-142) noted similarly that "Black people see the police department as the police of all the people, not just for white people.

That's a major achievement. The department more genuinely represents Detroit. This is symbolic of what Coleman has done for black people."

Documenting these changes in police behavior with harder evidence is difficult. Other observers, however, have made similar judgments about the transformation in police–community relations (Jones, 1978, p. 115), and, like the black officials in the two administrations, they attribute much of the responsibility to the two mayors.[2] To the extent that changes in police behavior have occurred, they are all the more striking given the fact that the police, particularly in Detroit, believed they bore the brunt of the costs of transition to black rule.

Some elements of mayoral policy toward the police were zero-sum in character, as far as whites were concerned: for example, the replacement of white police chiefs with black men, and the affirmative action policies regarding hiring and promotion discussed below. It is, of course, quite likely that much of the change in police behavior can be laid to the increase in the number of minority officers both on the street and in positions of authority, a specific policy goal of both black mayors. But the change in police behavior is also probably a result of such specific actions as the abolition of the STRESS unit in Detroit and of the more diffuse moral authority asserted by the black mayors. These must be seen ultimately not in zero-sum terms but as a consequential gain for both blacks and whites.

White interests, either at the elite or mass level, are not basically threatened by such changes. Black gains in terms of the respect they command from the police have not come at the expense of whites. Finally, it is important to note that the nature of police behavior has long been an issue of supreme importance in black ghettos all over America, and the mayoralty is far from irrelevant as a vantage point from which to affect it.

RELATIONS WITH THE FEDERAL GOVERNMENT.

Since the days of Mayor Richard Lee, whose small Connecticut city seemed at one point to have cornered the lion's share of federal urban renewal funds, it has been widely conceded that the character of the mayor can make a significant difference in what a city gets from Washington. The energetic pursuit by a mayor of good relations with the federal government can be translated directly into federal largesse. Thus

[2]James Q. Wilson (1968, p. 233) argues that the most important way in which a local political culture impinges on police behavior is through the selection of a police administrator and the molding of expectations that govern his role. In Atlanta and Detroit the mayors are the key figures in the selection process and may therefore take substantial responsibility for whatever changes ultimately occur in police practices.

the mayor's office provides a genuine base from which the activist mayor can exercise leadership in this area. If there is little formal power in the mayoralty with regard to federal–city relations, there are nevertheless substantial opportunities that mayors can generate and exploit to their city's advantage by cultivating links to Washington.

In losing their hold on city hall, white elites could no longer presume to control the forum of the mayoralty and the nexus of opportunities it provides for tapping the federal pipeline. But the black mayors in Detroit and Atlanta quickly demonstrated both that they could command attention in Washington and that the things they sought were of interest to white business as well as to the black community. By 1978 Detroit ranked first among the nation's 48 largest cities in the amount of federal money received as a proportion of revenues raised locally. For every dollar the city of Detroit raised itself, it was receiving 69.6¢ from the federal government, compared to an average among this group of cities of about 50¢ (Stanfield, 1978, pp. 868–869). Although Atlanta's receipts from Washington fell below this average, the city experienced a 140% increase in federal funds between 1976 and 1978, which placed it among the most active growth leaders (Allman, 1978, p. 47). Clearly, both cities were winning better than average shares of the federal pie toward the end of the 1970s, even in a period of general growth in federal–city fiscal transfers.

In a number of particular ways the two cities appeared to occupy an especially visible place in the federal scheme. Both black mayors claimed special access to President Carter, Jackson by virtue of his association with Carter in Georgia politics, and Young because of his extremely early and vigorous support of Carter's candidacy. Several close associates of the two mayors were in fact subsequently appointed to positions in federal agencies by the President, including Bill Beckham, Young's former deputy mayor who became Assistant Secretary of the Treasury Department. Other officials from Detroit and Atlanta took high positions in HUD, OMB, and the Department of Transportation.

Both mayors also maintained close relations with Secretary of Housing and Urban Development Patricia Harris. During a visit to Detroit in the fall of 1977 Secretary Harris made an emotion-laden tour with Coleman Young to the Blackbottom area, the neighborhood in which the mayor had grown up. On the spot, Harris committed HUD to $80 million in mortgage guarantees for 2100 units of housing in the area. A short time later a Detroit savings and loan association announced plans to develop housing in Blackbottom without the backing of the federal mortgage guarantees, marking the first time in 30 years that a privately developed housing project had been contemplated by white financial interests in the city.

Among the most recent visible benefits the two cities gained from the federal government was inclusion in a set of economic development programs backed principally by the Economic Development Administration. Detroit and Atlanta were among the initial 37 areas and localities included in the EDA's long-term demonstration economic development process, called Comprehensive Economic Development Strategy (initiated in 1978); both were also among the first 25 cities in an interagency program called the Neighborhood Business Revitalization Program (announced in 1978). In addition, Detroit was one of seven designated "Commerce Cities," which involved yet another pilot program (1978) for economic development. Finally, both Detroit and Atlanta were among the 45 cities and towns to receive Urban Development Action Grants in the first round of funding announced in April 1978. The grants, authorized by the 1977 act extending the Housing and Community Development Program, amounted to $1.2 million for Detroit and $1.7 million for Atlanta, and were designed, like the other economic development programs, to spur private business investment.

All of these Carter administration efforts were designed as pilot programs to create jobs by encouraging private investment in central cities through various subsidies and financing schemes. Federally guaranteed loans, financing assistance, comprehensive economic planning, and the training of local public officials in economic development strategies appealed to black politicians and public servants whose main concern was the creation of jobs, as well as to the white business communities in both cities. Officials believed that the inclusion of their cities in these pilot programs was a product both of their mayors' grant-seeking activities and good reputations in Washington, and of the business–public sector coalitions that both men seemed to have successfully fashioned.

ECONOMIC DEVELOPMENT

Subsidizing the profit-making capacities of private enterprise through the expenditure of public monies has been a traditional feature of American government. But the notion of a "partnership" between the public and private sectors to help "save" the cities achieved a renewed currency with the emergence of President Carter's urban program (see Report of the President's Urban and Regional Policy Group, 1978, p. 7).

Urban economic development, as it is understood by both white and black elites in Atlanta and Detroit, may be defined as the expenditure of public resources to encourage private investment for the purpose of creating or saving both central city jobs and tax base. Blacks in government and whites in the business community tended to view economic development strategies (in a nascent stage in the middle of 1978) as mutually beneficial. White interest in subsidized economic growth converged with the city governments' hopes of maintaining or expanding the local tax base and job opportunities. In both cities the bulk of the public money spent to "leverage" private investment came from the federal government. But matching local money and an elaborate local infrastructure to plan and administer the use of federal funds was a substantial local contribution to these efforts.

Although many of the resources for development were federal in origin, the initiative to seek such funds and develop local capabilities lay squarely with the mayors. This was particularly evident in Detroit. During Gribbs's administration the city bureaucracy had only two or three people assigned to industrial and commercial development tasks. Under Young this number had increased to nearly 30 by mid-1978. The number of city employees engaged in economic planning went from 1 to 12 in the same period.

State enabling legislation, pushed by the Young administration, allowed Detroit to establish an Economic Development Corporation (EDC), whose function is mainly to acquire and develop land parcels for industrial parks, and a Downtown Development Authority (DDA), which may formulate and implement plans to develop downtown commercial and residential real estate. In addition, the Detroit Economic Growth Council (DEGC), established by the mayor in 1976 to study the city's fiscal condition and composed of representatives from business, banking, and the public sector, was transformed into a permanent quasi-public body in 1978. With financing shared by the state, city, and local business,[3] the function of the new DEGC is to provide staff services to the EDC and the DDA and to channel money from the federal Economic Development Administration by advising the city (as recipient) on its placement. Federal commitments to job development, industrial expansion, public works, and residential rehabilitation in Detroit totalled $800 million between 1975 and early 1978.

Economic development activity, funded through EDA grants, money from the Small Business Administration, federal public works funds, and HUD grants is guided in Detroit by the mayor's "Move Detroit Forward" plan. The first version of this plan, developed early in Young's

[3]The city and the state each contribute $375,000 to the annual $1 million budget; private business contributes $250,000.

first term, is thought to have been influential in the formulation of President Carter's urban program (I-142, 145). A second version, taking account of the new state enabling legislation permitting the city to establish various development authorities, was released early in 1978.

The elaborate infrastructure for economic development in Detroit had not yet been matched in Atlanta in 1978, although various plans were on the verge of implementation. Jackson had established a Mayor's Office of Economic Development in 1977 to plan, review, and evaluate proposals for spending public money for job and business development. Plans were to be implemented through the new Atlanta Economic Development Corporation and a proposed Local Development Corporation designed to channel loans from the Small Business Administration. However, public works programs in Atlanta far outstripped Detroit efforts in this area, due mainly to the massive ($450 million) airport project, the largest capital development effort ever funded in the Southeast. Although some plans for this had been laid prior to Jackson's initial election, final planning and implementation were shaped by the Jackson administration.

In all of these various efforts public loan guarantees, local matching funds, public facilities improvements (e.g., parking ramps), outright grants, land acquisition, and so on were to be used to encourage private investment in the expectation that this would preserve and generate jobs and tax revenues. In 1978 it was too early to tell whether the mayors would succeed in these efforts, but in both cities black public officials and white business representatives spoke optimistically of multi-billion-dollar development schemes and the general prosperity they were expected to bring.

AFFIRMATIVE ACTION

Both black mayors actively sought to increase the number and proportion of blacks in all areas of city employment. Since total city employment decreased in Detroit and grew only slightly in Atlanta during the mid-1970s, black gains came to some degree at the expense of whites. White out-migration from the two cities, however, reduced to some undetermined extent the expected share of city jobs that whites would have gotten if such jobs were distributed on a strictly proportional basis. Thus each additional black city employee (assuming the total number of employees is held constant) did not displace a white counterpart.

In both cities the mayors asked black men and women to fill nearly half the appointive positions in city government (see Table 7.1). Blacks were more likely than whites to hold paid administrative positions as heads of departments, although whites were more numerous in both cities on the various unpaid boards and commissions, many of them

TABLE 7.1

Racial Breakdown of Mayoral Appointments in Detroit and Atlanta

	Atlanta[a]		Detroit[b]	
	White	Black	White	Black
Department and agency directors	12	15	19	20
Boards and commissions	252	237	71	56
Mayor's executive staff	unavailable		9	12
Other (judicial agencies, task forces)	2	5	48	20
Race unknown or other[c]	55		43	
Total appointments	578		298	
Percentage black of those whose race is known	49%		42%	

Source: Data supplied by mayors' offices.

[a]Figures for Atlanta are cumulative for the 1973–1977 period.

[b]Figures are for 1977 only.

[c]This category includes a few Orientals, Hispanics, and Native Americans, as well as a number of people whose races were unknown to officials compiling these data.

honorific in nature.[4] Although black representation in all appointive positions prior to 1973 can only be very roughly estimated, the number of blacks in those years did not come close to reflecting their proportion in the city population.

Affirmative action at the appointive level was seen as important not only to ensure the representation of black interests but also to influence the structure of job opportunities for blacks within city agencies. Personnel directors in both cities (both of them black women in 1978) stressed that the evaluation and promotion of employees are the ultimate responsibility of department heads. Both believed that sympathetic black administrators had favorably affected the distribution of blacks in responsible positions (I-143, 238, 243), and affirmative action data appear to provide evidence for these claims. Table 7.2, Part B, shows striking increases in both cities between 1973 and 1978 in black representation in managerial civil service positions. These in-

[4]Daniel P. Moynihan and James Q. Wilson's study (1964, p. 299) of patronage politics in New York State found that out-groups (primarily of upper-class Protestant background) during Democratic administrations tended to be appointed to honorific boards and commissions, whereas Catholic, Jewish, and racial minorities tended to be appointed to paid administrative positions.

TABLE 7.2
Affirmative Action in City Government Employment

	1973	1974	1975	1976	1977	1978
A. Percentage minority in total classified service						
Detroit	45.1	NA[a]	NA	NA	53.1	53.9
Atlanta	41.5	42	46	47.5	49	55.6
U.S. cities [b]	17.9	19	20.6	NA	NA	NA
B. Percentage minority in managerial and professional positions in classified service						
Detroit						
Managerial grade	12.1	NA	NA	NA	23.5	32.2
Professional positions	22.8	NA	NA	NA	41.1	37.6
Atlanta						
Managerial grade	NA	NA	NA	NA	NA	32.6
Professional positions[c]	19	20	23	25	28	42.2

Sources: Detroit and Atlanta figures were supplied by the respective City Personnel Departments. National totals are contained in U.S. Equal Employment Opportunity Commission, *Minorities and Women in State and Local Government* (Washington, D.C.).

[a] NA = not available.

[b] All cities in SMSAs. Blacks made up 22.3% of the population in these cities in 1974.

[c] Defined by job category, for example, engineer, budget analyst, architect.

creases occurred at a faster rate in fact than the increase in overall black city employment, as a comparison of Parts A and B in Table 7.2 indicates.

Affirmative action efforts in Atlanta and Detroit predate the election of the black mayors. Mayor Gribbs had ordered all Detroit city departments to develop affirmative action plans, and his order was supported by a city council resolution in 1971. In Atlanta Mayor Massell had created the post of affirmative action officer in the Personnel Department in 1973. Furthermore, in 1972 the U.S. Congress extended the equal employment opportunity clause (Title VII) of the 1964 Civil Rights Act to include state and local government. One effect of this was to force local governments to make annual reports to the Equal Employment Opportunity Commission on affirmative action gains.

Despite these early actions, however, the influence of the black mayors seems clear. The Atlanta affirmative action post, though created in 1973, was not filled until early 1975 by Maynard Jackson. Both cities initiated extensive and active recruitment efforts under the black mayors to search out minority job candidates. During the black mayors' first terms both cities had reevaluated and changed their selection procedures to place less emphasis on standardized written exams. Atlanta had instituted an extensive employee counseling program and an internal discrimination complaint system. Detroit's Personnel Office had begun to identify promising minority professionals still in college in order to lure them into city government with the offer of internships during the last year of their college training. And, as we have seen, the appointment of black supervisors and department heads apparently had a ripple effect, opening advancement opportunities for blacks farther down in the employment hierarchy.

A study of black employment in local government prepared by the Southern Regional Council also suggests, by implication, the importance of the presence of black leadership. A large black presence in a city's population is not enough to ensure substantial black employment opportunities. Examining employment patterns in 16 southern cities, which averaged more than one-third black in population, the Council found that blacks and women together represented only 7% of all employees earning over $13,000 in 1975. None of these cities had a black mayor. Only 2 of the 16 cities had affirmative action plans with specific goals for minority and female hiring (NYT, 25 May 1978).

Affirmative action efforts in Detroit and Atlanta also extended to the police forces. During 1966–1967, blacks constituted only 9% of Atlanta's police force, 5% of Detroit's (Fogelson, 1977, p. 248). By 1978, 33% of Atlanta's police were black, as were about 30% of Detroit's. The increase in blacks on the force, an explicit goal of both mayors, is not simply a function of the greater proportion of blacks in the population. Estimating the black population at roughly 40% in 1967 in both cities, the underrepresentation ratios of black police to black population for Atlanta and Detroit were about .23 and .13, respectively.[5] By 1978, assuming black populations of roughly 55%, underrepresentation ratios had diminished to .60 in Atlanta and .55 in Detroit.

[5]The ratio of underrepresentation is calculated by dividing the percentage of blacks on the police force by the percentage of blacks in the population. A score of 1.0 indicates proportional representation. A score of 0 indicates no blacks on the force.

Atlanta's public safety director was black as well as two deputy directors during the 1973–1978 period. Detroit's chief of police was black, and blacks occupied 9 of the 21 district commander positions in 1978.

Most black city administrations have not been content to rely solely on affirmative action plans to increase black representation among local government employees. They have, in addition, sought to structure the recruitment pool from which city employees are drawn to favor blacks through the imposition of a residency requirement.[6] With the city no longer obligated to draw from a predominantly white metropolitan labor pool, black job aspirants in the increasingly black central city presumably find their chances of landing a city job increased.

Detroit has had a residency requirement for all city employees, both appointed and classified personnel, since the turn of the century. During Young's tenure the city responded vigorously to a challenge to the residency rule launched by the Detroit Police Officers Association in 1975 during contract negotiations with the city. The ordinance was upheld in binding arbitration.

In Atlanta the current residency requirement was apparently a special project of Maynard Jackson's. Although the city had once had a stringent requirement applying to all city personnel, it had lapsed just prior to World War II. As vice mayor, Jackson began to study the possibility of reinstituting such a law. At that time nearly 90% of all white policemen lived outside the city. When he became mayor, Jackson immediately imposed by executive order a residency requirement on all appointed personnel. In addition he pushed for a city ordinance applying to policemen and firemen, which the city council passed in 1976. It applied, however, only to new hiring.[7]

CITY PURCHASING AND PRIVATE SECTOR HIRING PRACTICES

When the black mayors came into office in Atlanta and Detroit, only a tiny percentage of city purchases and contracts were made with firms owned by blacks. By 1978, however, both cities were spending one-third of their purchasing dollars with minority firms (see Table 7.3). Although the two cities let contracts on the basis of competitive bidding for every-

[6]Residency requirements that restrict city jobs to those who live within the city limits are not designed solely to favor minority groups. Approximately three-quarters of all cities over 250,000 impose such requirements on their police. Data on the incidence of more general residency laws are not available (ICMA, 1974, p. 222). Nevertheless, it is noteworthy that a number of large black-mayor cities have actually instituted some sort of residency law during the black mayor regimes. These include Atlanta, Newark, Gary, and Washington, D.C.

[7]Just as the ordinance passed, the courts imposed a hiring freeze on the police and fire departments in which white-instituted discrimination suits were being settled. Two years later the freeze had not been lifted. Thus, the residency requirement had not yet had any effects in 1978.

TABLE 7.3
Percentage of City Business[a] Going to Minority-Owned Firms

	1972	1973	1974	1976	1977[b]	1978[c]
Detroit	1.5	3	—	9[d]	20	33
Atlanta	—	2	13	—	16	33[e]

Source: All figures were supplied by the respective city purchasing departments.

[a] City business includes supplies, equipment, professional services, contracts.

[b] Atlanta total spending in 1977 was $227 million. Detroit spending amounted to $108 million. The lower figure for Detroit may be accounted for in part by the fact that substantial spending— for repairs and maintenance, for example—is handled by individual departments and does not come through central purchasing. Minority figures for those expenditures are not available.

[c] 1978 figures for first quarter only.

[d] First full year of operation of Ordinance 52-H.

[e] First quarter after passage of Finley Ordinance.

thing from paper clips and janitorial supplies to major public works construction, both have developed a number of ways to ensure substantial minority participation in city business.

In Atlanta the so-called Finley Ordinance passed during Jackson's first term establishes a post of city contract compliance officer. This official may set minority hiring goals for all firms planning to bid on each purchase or project. Firms that cannot or will not attempt to meet the target established by the city are declared ineligible to bid. Firms not in compliance but with a plan to hire minorities to meet the city's goals may bid, but their bid may be turned back, even if it is the lowest, if the contract compliance officer is not satisfied that it is a good faith plan.[8]

Since many white firms could not begin to meet minority hiring goals, which range as high as 40% on some construction projects, the city allows joint ventures. These are arrangements in which a white-owned firm and a black-owned firm merge their resources, submit a joint bid, and share the contract and profits on the basis of a negotiated formula. Thus small black firms with predominantly black employees enable many larger, noncompliant white firms to participate in city contracts. But the small black firms, which could not have competed on their own, clearly benefit from the arrangement.

Although a number of business leaders (particularly in construction) have complained vociferously about joint-venturing and the policy of city attempts to influence hiring practices, only one lawsuit had been filed by mid-1978 to challenge the ordinance. Despite complaints, few

[8]The plan is similar to one put into effect at the federal level. In 1965 President Johnson ordered all firms with more than 50 employees doing at least $10,000 worth of business with the federal government to submit affirmative action plans to Washington. Failure to comply was to result in exclusion from further bidding. President Carter, however, is the first president to enforce this executive order vigorously (NYT, 26 May 1978).

firms had actually withdrawn from competitive bidding on city work. Of some 3000 firms doing regular business with the Atlanta city government, only 30 were on the ineligible list in mid-1978. Many of these latter were not local firms. Several major companies, however, mainly in the construction industry, had never sought to comply with city hiring targets and therefore did not bid on public sector contracts.

Although Detroit also requires affirmative action reports from all firms submitting bids for city work, it has gone a step further than Atlanta. Ordinance 52-H, passed in 1975, establishes a preference system for local firms:

> [In comparing bids] the bid of any Detroit-based firm shall be deemed a better bid than the bid of any competing firm which is not Detroit-based, whenever the bid of such competing firm shall be less than two percent lower than the bid of the Detroit-based firm on any contract bid upon in an amount of $100,000 or less and less than one percent lower on any amount bid in excess thereof.

The ordinance is specifically designed to aid "those small business concerns which . . . are owned by socially or economically disadvantaged persons," according to its lengthy preamble.

In addition to preferential treatment of Detroit firms—which are not all minority-owned, of course—the city Purchasing Department actively solicits bids from minority firms. The department also runs a vigorous advertising program aimed at minority entrepreneurs: "We encourage them to come in and become aware of what products we buy," an official in Purchasing (I-140) explained. "Often businesses aren't aware of the diversity of the city's purchases."

By compelling firms interested in bidding on city business to meet affirmative action criteria, the city possesses a modest tool to influence private sector hiring patterns. In both cities officials assert that elite level political support for affirmative action hiring increased under the black mayors, although these assessments were made before the 1978 Supreme Court ruling in the Bakke case (I-145, 244). Both cities broaden their influence over private sector hiring by requiring affirmative action on the part of any firm benefitting from the expenditure of public funds apart from the purchasing and contracting process. For example, Detroit's Economic Development Corporation writes affirmative action target goals into a contract with a firm that plans to locate on industrial park land that the EDC has assembled and developed.[9] In addition the mayors

---

[9]All contracts financed primarily with federal money—public works, mass transit funds, economic development—also carry with them requirements that minority firms get a certain proportion of the business and that minority workers hold a certain percentage of the jobs.

have used the moral authority of their positions to speak out against discrimination in corporate boardrooms and executive offices. Maynard Jackson's pressure is credited with gaining token black representation on one of the major Atlanta bank boards.

CONCLUSION

To suggest that city government is a hollow prize for blacks, just as they have begun to win important mayoralties, is a curious conceit of the cynic. Few of those same observers suggest that it is an irrelevant goal for white aspirants. Despite the well-acknowledged limitations of city hall, the office has not been found to be irrelevant to certain meaningful interests of both blacks and whites in Atlanta and Detroit. Of course the power and influence inherent in the mayoralty must be placed in perspective. But as Leonard Cole (1976) has written, "Whatever the shortcomings of the American political system, the argument that local officials are unimportant is specious. They do face red tape and bureaucratic roadblocks; they are often sandwiched between an abusive electorate and unconcerned state and federal officials. . . . But at the same time there *is* power within local government [pp. 23–24]."

What, in sum, is the nature of that power and how does it bear on the question of white acquiescence to black rule?

Transition was not in certain ways cost-free to white elites. They could no longer assume routine access to the decision-making process, nor could they expect in the long run to control the forum for leadership that the mayoralty affords. In more concrete terms, the possibility of their using the police as a means of repressing the black underclass has greatly diminished, since that instrument of internal social control increasingly rests in the hands of blacks themselves, from the street level to the police chief's office. In addition, white business discovered precipitously that it would have to share its former monopoly over city business contracts with black-owned enterprises, and that access to that business and to certain other benefits derived from the expenditure or backing of public monies would depend on more regulated hiring practices. However, the most visible "costs" to the white community, namely affirmative action in public sector employment, were borne not so much by white elites but by whites in the labor force. Nevertheless, the bitterness at preferential hiring and promotion, felt most acutely perhaps by the police, was not translated into effective political opposition to black rule in either city.

White elites quickly came to see, of course, that transition to black rule did not affect certain areas of white interest at all, and in some instances produced benefits that the black mayors believed were important to both races. As long as business activity was dependent strictly on private sector financing and contracts, city government (or Washington, for that matter) could do little to elicit socially responsive investment or hiring decisions. (For example, although such situations are increasingly rare, several major construction firms in Atlanta still were not hiring many blacks in 1978 and were able to maintain that policy because they restricted their activity entirely to the private sector.) However, the incentive of public economic development money generated in large part by mayoral activity is likely to bring more and more firms within the orbit of affirmative action programs, but at the same time make continued commitment to the city in the form of investment an economically attractive option. Both business and city hall believed that all stood to gain from such economic development policies.

John Mollenkopf (1978, p. 146) has suggested that the black mayors are little different from their white predecessors in that they too have sought to develop and lead "progrowth" coalitions. These are the urban renewal and development groups of the 1950s and 1960s—developers, bankers, city hall, and the Chamber of Commerce—reincarnated but with a different racial mix. Although the point is to a limited extent valid, there are some crucial differences.

The old progrowth coalitions concentrated on downtown development, "slum" clearance, and highways to bring commuters to the city core. The new progrowth coalitions have sought to build on vacant or abandoned land (Detroit's riverfront development and Atlanta's Bedford-Pine area are cases in point); included neighborhood rehabilitation, small business revival, and housing as important components of their development schemes; pressed private industry on affirmative action as a condition of their participation; and generally opposed destructive highway corridors. Most important, the new coalitions have been built on the basis of a perceived mutuality. Black interests, *as they are defined by the black administrations*, have played as central a role as the interests of white capital in the establishment of economic development goals. Indeed, unlike the urban renewal groups of Allen's Atlanta or Cobo and Cavanagh's Detroit, the new coalitions represent an attempt by blacks to harness white economic power rather than vice versa. It is too early, perhaps, to evaluate the ultimate success of these efforts by the black mayors for the ordinary people of their cities. But it is clear that blacks are no longer the weak adjuncts to the dominant white coalitions of prior decades, standing helplessly by, watching the men they helped to elect encircle their ghettos with office towers and freeways.

In summary, the notion that whites acquiesced so easily to black rule because the mayoralty is irrelevant to their interests is vastly over-simplified. It is more accurate to argue that the evident costs of transition for white elites have been balanced out by benefits to some degree. Yet it cannot be said that the black mayors bought off the white business community, for the benefits whites derived (or hope to derive) from economic development particularly are of equal interest to the black administrations. Nor can it be argued that white elites could afford a certain tolerance because black government is so dependent on white business investment and commitment. For one thing, many of the accomplishments of the black mayors were achieved without the substantial support of whites in the business community. For another thing, it is clear that dependency is increasingly a two-way relationship, particularly insofar as public funds for private economic development must be solicited and administered by the city. It seems evident, then, that the benign quality of the white response cannot be explained very well as a function of the limitations or the irrelevance of the mayoralty.

8

Explaining the White Response to Transition: Environmental Factors

An explanation of white acquiescence to transition cannot be built on the false assumption that white interests are autonomous from local politics. A more convincing explanation for the character of the white response focuses on a variety of environmental forces at work in the society and in the particular cities in question. White response may be viewed as involving several kinds of adjustments at both the psychological and strategic levels, and different sets of environmental factors may be brought to bear for each of these adjustments.

Why were white elites willing to regard black rule as a legitimate outcome of the political process in the first place? An examination of the cultural environment in which transition took place may elucidate the reasons for the peaceful character of the adjustment, set against national and local histories of racial antipathy and specifically local fears of black government. By the term *culture* I mean here those shared mental products that give shape and purpose to social experience; thus, certain attitudes, norms of political conduct, standards, aspirations, and traditions and myths that relate to the social collectivity and are widely shared may be considered aspects of culture.

However, not only did white elites respond to transition by accepting the principle and fact of black rule, but they also appeared to maintain a certain level of commitment to their cities. There was, for example, no evidence of withdrawal of economic resources or civic energies. Indeed, just the opposite was found: a pattern of continued investment, a commitment to the city and its government filtered through the prism of business values, and participation in public service enterprises. Thus another explanation for white acquiescence to transition is the compelling nature of certain economic stakes.

But neither acceptance of the legitimacy of black rule nor the maintenance of commitment to their city precludes the possibility of political opposition. Yet, as we have seen, such a response to the black mayors was slow to form, fragmented in character, and spiritless. Although some white elites genuinely liked the black mayor of their city or set out on a deliberately cooperative course as the most sensible way of protecting their interests or those of the city, opposition was made difficult in any case by certain characteristics of the sociopolitical structure of the two cities that served to undercut and diffuse opposition.

In sum, the environmental explanation suggests that the nature of white elite adjustment to transition in these particular cases is in large part a function of both national and local cultural elements, the economic environment, and structural features of the two cities.

THE CULTURAL IMPETUS FOR ACQUIESCENCE

A complex of forces, both national and local in origin, has shaped the cultural context of the politics of Atlanta and Detroit in such a way as to permit the softening in the 1970s of the sharp edges of racial tension, particularly in the area of elite interactions. At the most general level the impact of culture on mayoral politics in the two cities originates in the complex of normative presuppositions that Americans tend to bring with them to the political arena, particularly those regarding elections and the conduct of candidates. If Americans are no strangers to occasional electoral corruption, they nevertheless overwhelmingly accord elections a degree of legitimacy that virtually precludes the resort to extralegal means to overturn electoral results and helps losers to accept the decisiveness of the outcome. Concession speeches of defeated candidates reflect these norms to the extent that they are filled with promises to cooperate now with the victor but to return to fight another day. If losers believe that the majority is not always right, they seldom doubt that it has spoken clearly and that its voice, even if misguided, must be heeded in the leadership selection process. Implicit in these norms of fair play is the notion that personal political ambition must to some degree be limited by the requirements of the larger society for stability and order. Americans have no historical precedents in domestic politics for using violence to overturn electoral results.

We saw in Chapter 4 the widespread acknowledgement in both cities of the existence of a black majority and its legitimate right to elect one of its own to office. This is clearly a product of these "fair play" electoral norms. We may also observe other manifestations of this complex, particularly in Detroit. Witness the reflections of John Nichols (17 March 1976), whom Young defeated in the mayoral race of 1973:

> Coleman and I realized in the campaign that we didn't want to inherit a city torn by racial disorder. Coleman told me a few months ago, *"We could have torn this city up."* But we didn't. We both maintained our dignity in the face of some assaults by the opposition.

And on his loss Nichols commented, "The fact that I lost didn't send me into my tent to sulk." These remarks illustrate not only the acceptance of defeat that the society generally expects of losing political contestants but also the sense that a larger responsibility must constrain the individual competitors in their quest for office.

Electoral competition may on occasion be too bitter to bring into play the expected public gestures of acquiescence and cooperation, but the changeover to black rule in Atlanta and Detroit was marked by cooperative rituals. These were limited in Atlanta mostly to such symbolic signals as Massell's attendance at Jackson's inaugural and the spate of newspaper editorials encouraging the new mayor. But Massell (18 June 1975) also exhorted the young business leaders of the community, by his own report, "to help Jackson become the greatest mayor in Atlanta history. Help him, join him, I told them." Although the outgoing mayor in Detroit in 1973 was not the defeated candidate, his approach to the actual transfer of government is illustrative of the fair play norms at work.

> When Coleman took over, I met with him four times. I wanted to do for him what Jerry [Cavanagh] hadn't done for me. I met with departmental heads in Coleman's presence. It was important that city whites accept the first black mayor. His people asked me—sheepishly—to attend his inauguration. I said sure. He's my mayor. It was important with the first black mayor [Gribbs, 17 March 1976].

The fair play norms—the belief in the legitimacy and decisiveness of elections, the display of (at least) ritual gestures of cooperation and conciliation, the acceptance of defeat—could be brought so easily into play in Atlanta and Detroit in part because a variety of other cultural influences converged in such a way as to support the application of these norms to the situation of transition. For one thing the rise of black majorities and black influence in these cities must be viewed against a 30-year period of steadily increasing levels of racial tolerance nationally among whites.[1] National longitudinal opinion data show that support for segregation of the races has diminished to the point where it is no longer an issue in certain areas of social life (for example, in transportation and public accommodations). Attitudes also appear to be softening in regard to black political aspirations.[2] In a study set in New Jersey, for example, Leonard Cole (1977, p. 309) found that most whites in his 1972 survey agreed that it was important in their city for blacks to hold elective office in order to represent black interests.

There is some modest evidence to indicate that neither of the cities in question in this study was immune to the forces producing these opinion changes. When Arthur Kornhauser (1952, p. 100) surveyed Detroit residents in 1951, he found that 68% of the whites in his sample supported racial segregation in one form or another. Twenty years later, however, Aberbach and Walker (1973, p. 154) found that the proportion of whites in Detroit that supported segregation had diminished to 20%. Comparable survey data do not exist for Atlanta, but Jack Walker (1963, p. 230) was able to describe Atlanta in 1960 as still "basically segregated." Yet a 1968 survey of racial attitudes in that city concluded that only one-quarter of the white population could be described as "racist," that is, as opposing all efforts to end discrimination and demonstrating hostility toward blacks (Hutcheson, 1973, p. 42).

Other, primarily locally generated cultural forces have added their cumulative impact on the ways in which white elites have adjusted to transition. These include the union legacy in Detroit, a reinterpreted "New South" myth in Atlanta, a brief but significant history of experience with black political officials in both cities, and the sensitivity of the new black mayors to the fears of their white constituents.

Perhaps the major institutional force for progressive race relations in Detroit has been the United Auto Workers union. It is significant that the first act taken by Leonard Woodcock upon his election to the presidency of the union in 1970 was to fly to Georgia to join the Southern Christian Leadership Conference Poor People's March (Reuther, 1976, p. 471). Although elites in Detroit were careful to distinguish between the racial views of the union's leadership and its white rank and file,

[1] For summaries of such data, see Greeley and Sheatsley (1971) and Erskine (1973).

[2] For trend data for the period between 1956 and 1976 showing increasing willingness among whites to vote for a "qualified" black for president, see Dennis (1976, figure 1).

they nevertheless argued that the UAW could take major responsibility for the ease of transition at the elite level by having sought to create a climate of racial accommodation and by supporting black political and social aspirations (I-14, 15, 112, 114, 119, 138). Although the UAW must bear some responsibility for discriminatory labor practices in the auto plants in the 15 years after World War II (Widick, 1972, p. 149), the union was nevertheless actively supporting racially liberal mayoral candidates (albeit unsuccessfully) all through this period (Gray and Greenstone, 1961, p. 372). The role of the union as a force for amicable race relations was clearly illustrated in the comments of one UAW official (I-138): "This is the only city in the country that puts together an annual NAACP Freedom Fund dinner. It's a $100 a couple deal. . . . I think the reason it takes place is the UAW."

If the progressive tradition of the union in Detroit has served as a persistent stimulus for better race relations, then a version of the New South notion has performed a similar, if perhaps less forceful, function in Atlanta. The New South "creed" developed after the Civil War, giving rise to both a movement and a program designed to bring the South into the mainstream of the nation. One of the major shapers during the 1880s of this myth of purgation and rebirth was an Atlantan, Henry Grady, whose version of the new order stressed racial peace and modern economy based on industrial development and "scientific" agriculture (Gaston, 1970, p. 7). In the succeeding century, the central tenets of the creed were reinterpreted to suit the character of the times. At one point the separate-but-equal doctrine was understood to be a just means to the realization of harmonious race relations, but as early as the 1940s the Southern Regional Council was calling for an end to all race discrimination in its magazine *New South*.

The troubles in the lunchrooms, courthouses, and bus stations of the South in the early 1960s stayed talk of a New South, but by the middle of the decade there was a revival of the notion. Observers saw the New South upon the nation at last, a product of a surging economy and population growth, gains in black political influence, and the rise of the "new" southern politicians. Governor Jimmy Carter of Georgia exemplified the tone of the revival by pledging from the state house in Atlanta that "the time for racial discrimination is over [Dietsch, 1971, p. 617]."[3] The contribution to the revival by Atlanta elites was to begin to reinterpret the creed in such a way as to link racial peace with economic prosperity, seeing the latter as dependent on the former.

At the beginning of the 1960s Atlanta leaders were convinced that a

[3]See also Eggler (1970) and Nordheimer (1974).

repetition of Little Rock's agony of the 1950s would bring economic disaster to the city (Allen, 1971, pp. 52–53). The Atlanta Chamber of Commerce was just about to launch a $1.5 million public relations campaign to spur the local economy (the second "Forward Atlanta" program) that would make Atlanta the nation's most advertised city in the next 5 years. To protect their investment in this campaign and the city's economic future, business and political leaders decided to support school desegregation in 1961. The effort was led by Ralph McGill, editor of the *Constitution*. Police Chief Herbert Jenkins and Mayor William Hartsfield stressed Atlanta's determination to accomplish desegregation peacefully. Mayor Allen later became the only southern mayor to testify before Congress on behalf of the 1964 Civil Rights Act, having previously helped to encourage the desegregation of some of Atlanta's public facilities before the law required it. On the basis of such efforts Atlanta earned a national reputation as "the leader of the New South" (Hein, 1972, p. 207).

The reputation of the city as a "pacesetter" and as "too busy to hate" has continued to be important to Atlanta's white elites (Hartshorn *et al.*, 1976, pp. 9–10). In the minds of the city's businessmen, good race relations, favorable image, and prosperity are all inextricably linked in an unbroken chain (Hein, 1972, p. 208). It would be incorrect, however, to attribute the effort to project an image of progressive race relations entirely to economic motives alone. Many of Atlanta's white civic leaders were proud of their city's reputation as a leader among cities in the area of race, and they acknowledged and accepted the moral burden of their position.

The accommodationist impulses encouraged by the UAW tradition of progressivism and by the responsibilities imposed by the reinterpreted New South creed in Atlanta were reinforced by certain experiences with black political power. Neither city, for example, was entirely unfamiliar with black officials before the accession of the black mayors, as we saw in Chapter 3. Atlanta had a black alderman by 1965 and Jackson himself had served as vice mayor from 1969 to 1973. In the mid-1960s Detroit's contingent of black officials included 3 judges, 10 state legislators, 2 U.S. Congressmen, and 1 city councilman.

Although the pioneer black officials, particularly the city councilmen, attracted substantial press attention in the two cities when they were first elected, whites quickly became accustomed to their presence in government. Among other things, whites learned that the preoccupations of office of black officials were to a large degree much like those of their white counterparts. If whites were still afraid at first of a black mayor, they were nevertheless prepared to learn (as they had already done with prior black politicians) that Jackson and Young were not intent on subverting the political order but on governing as best they could in a biracial setting (Tyson, 1975, p. 237). Both mayors facilitated this learning process by observing the norms that govern a victor's behavior toward losers in American politics (I-17, 236), and by their understanding of the novelty and difficulty of the white situation. "There is a peculiar anxiety which probably nobody was prepared to deal with, black or white," Maynard Jackson once commented ("Can Atlanta Succeed?" 1975). "That's the anxiety which must attach to a community which for the first time is a minority community [p. 41]." And Young noted a year after he had taken office for the first time that "many white people have learned that I don't have horns. Those who expected the worst—a black takeover—have found that that has not happened. The staff has been balanced. There's been no purge [*DFP*, 27 Jan. 1975]."[4]

The effect of this multitude of cultural forces has been to broaden the domain of the larger culture of accommodation to include racial politics. Although mainstream American politics have always for the most part been governed by a culture of accommodation, the inclusion of the racial component makes its form in Atlanta and Detroit comparatively new. That is to say, the imperatives and normative standards of that culture now apply specifically to a situation of black–white competition through established political processes. Notions of fair play now extend to blacks in a way they might not have as recently as the mid-1960s.

The culture of accommodation does not necessarily stifle racial conflict, discontent, or white unease. Many whites would surely prefer strongly to have a white mayor. But it does moderate the strategies in pursuit of conflict, virtually banish racism from the lexicon of political discourse, and place high premiums on cooperation and compromise as outcomes of racial conflict.

AN ECONOMIC EXPLANATION FOR THE MAINTENANCE OF WHITE ELITE COMMITMENT

For the majority of major economic actors currently in the city, abandoning the central business district for the greener pastures of suburbia or the sunbelt states is a chimerical option. Although departure may be both economically feasible and advantageous for small-scale operators or compelling for other reasons for individual homeowners, many of the major employers and tax producers consider the prospect of

[4]See also Tyson (1975, p. 238).

flight too costly to contemplate seriously. For the governors of the city it is this major economic sector, on balance, that provides the tax revenues, the investment, the jobs, and the civic leadership most worth retaining. In Atlanta and Detroit there is evidence to suggest that economic interests militate strongly against wholesale flight. It follows that if major economic actors find it in their interest to stay in the central city, then they will do what they can—particularly given current notions of corporate civic responsibility—to enhance the health and quality of their work environment.

Decisions to move from or stay in a city are normally taken in response to the costs and quality of the labor force, tax rates, insurance costs, the condition and size of existing physical plant facilities, and so on (see, e.g., Hamer, 1973, p. 10). But business people are becoming conscious of several new factors that may begin to slow metropolitan centrifugalism. These include the growing cost and scarcity of suburban land within a reasonable distance of the downtown and the increasing cost of transportation (I-118). In addition many cities, including Detroit and Atlanta, are witnessing revivals of old neighborhoods and their transformation into middle-class residential areas not far from the downtown, signalling a movement, even if presently small, of the return of a white-collar labor force from the suburbs.

But these developments, even if they are truly in motion, have perhaps not yet matured to the extent that they play a major role in decisions to stay in the central city. At present they seem to serve as reinforcing factors for other more immediate motives for staying put. In Detroit and Atlanta the most important of these motives is simply the maintenance and protection of the massive investment in bricks and mortar made primarily during the building booms of the 1960s and 1970s. Although figures on the nature of the building investment in the two cities are difficult to come by, and what are available are not always provided for comparable time periods, a picture nevertheless emerges of a massive development in both places of new physical facilities that cannot easily be left or sold.

During Coleman Young's first term the estimated value of new central city private investment completed and committed totalled nearly $2 billion. One billion dollars of new construction was actually under way or just finished. An additional $500 million in new construction had been committed, and private firms had invested in $300 million worth of new plant equipment (Detroit Mayor's Office, 1977, p. 4). In the years immediately preceding Young's election, Detroit, like many other cities, had undergone an office tower boom. Despite the completion of the massive Renaissance Center, however, demand for office space had virtually caught up with the supply, making the ownership of such downtown properties a profitable business. Not only was the Renaissance Center fully rented, but older office space was 80–85% full in 1978, an occupancy rate that had risen nearly 20 percentage points in 2 or 3 years (I-142).

In Atlanta estimates of the value of construction in progress when Maynard Jackson assumed office amounted to between $1 billion and $1.5 billion (AIA Journal, 1975, p. 34). In the decade prior to Jackson's election at least 17 skyscrapers of 15 stories or more were completed in Atlanta (Hartshorn et al., 1976, p. 14). In 1977 three new hotels opened within weeks of one another, including the 70-story Peachtree Plaza. During the 1960s over 12 million square feet of office space were built in Atlanta, and an additional 3 million were under construction in the first 3 years of the 1970s (Chamber of Commerce, 1974). A spokesman for Atlanta's Portman Associates (I-222), an architecture and real estate firm, noted in response to questions concerning the strength of his firm's commitment to the central city, "We're the largest single property tax payer in the city—about 10 percent of the taxable value of the city. This includes the Regency Hyatt. . . . You can't move a big hotel or office building." And an Atlanta Chamber of Commerce official (I-234) said simply in answering the same questions regarding business commitment to the city, "It would be impossible to abandon the downtown."

Not all of these investments were thriving in the first half of the 1970s. Atlanta suffered from an oversupply of office space and had warehouses unrented. Both cities lost a major hotel in the same period. But business leaders and economic analysts were nevertheless inclined to see investments in central city property as sound (Hartshorn et al., 1976, p. 42), and some even viewed the downtown as underexploited. A top official of Detroit Renaissance (I-124) observed in regard to the Renaissance Center complex, "The money people . . . felt they'd spread the metropolitan area out, but now they could revitalize the downtown. And they felt they could make a buck."

Besides the investment in physical plant facilities in the core city, incentives for maintaining a commitment to the center are also a function of a variety of well-established local habits and patterns adapted to the geography and sociometry of an urban hub configuration. A report by a long-time city planner in Atlanta (Hammer, 1974) stresses the dependence of the metropolitan region on that central city:

> Of all the metropolitan economies in the nation (with the exception of the Nation's Capital, Washington), Atlanta's is the most heavily based upon its central economic enclave. It is the core economy that supports

Atlanta's major regional and national function. It is the core economy that generates the broad range of support functions where most of the new job opportunities are created. . . . The city of Atlanta accounts for the bulk of the region's economic investments. All segments of the metropolitan population, white and black, inside the city and out, have an immense stake in the city's economic structure [p. 24].

The downtown, as a Detroit Chamber official (I-110) observed, "is still the center of action here. Some of the people who've moved out know that and regret their move." An executive in the Burroughs Corporation (I-132), headquartered in Detroit, commented, "Our business is business. We sell to businesses and banks especially. Financial institutions are in core cities. So we ought to stay close to our customer base." Other executives cited standard advantages of a downtown location such as the proximity of good hotel accommodations, the ease of face-to-face communication with other business people, and the availability of large quantities of top quality office space.

In both Georgia and Michigan state banking laws are such that a central location is made virtually mandatory for financial institutions, ensuring the continued presence of major banks in the main business district. In both states banks are allowed to establish branches only within a certain radius of the main office. By locating at the population and business center, banks can maximize the potential customer base to which both the main office and branches are available.

Atlanta and Detroit have also been—and continue to be into the future—popular convention sites. The convention business is an extremely competitive one: In 1977 the top 30 American and Canadian convention cities shared only 45% of all conventions held in the two countries (World Convention Dates, 1977). Chicago led all cities with 3.3% of all conventions scheduled. Atlanta was fourth with 2.6% and Detroit was eleventh with 1.7%. Both cities had slightly larger shares in 1977 of conventions scheduled for the future: Atlanta had 3% of all future bookings and Detroit claimed 1.8%. Besides helping to stabilize future demand at a high level for central city hotel space, such figures also ensure high future use of associated support and entertainment enterprises, ranging from restaurants and night clubs to downtown retail establishments. In response to a question regarding the possibility of moving their establishment out of the central city, for example, the public relations officer of a major Atlanta department store (Letter to the author, 18 Oct. 1977) answered that it had never been considered because convention business is concentrated downtown. Indeed, national statistics indicate that downtown retail establishments may count on sales averaging approximately $17.50 per delegate per convention (NYT,

25 Jan. 1976). Such projections support the conclusion of Hartshorn and his associates (1976) regarding Atlanta that "the central city retains qualities that assure not only survival, but continued vitality [p. 42]."

In seeking to protect their investments in the downtown and to maintain established patterns of business in the core city, white elites are not unaware of the role a black mayor might play as a guarantor of the social peace required to maintain the economic viability of a black majority city. The black mayors were seen by some to represent a hedge against the disruption of racial turmoil, although this was not a major theme in the interviews. Despite the fact that neither Detroit nor Atlanta escaped racial troubles during the black mayors' first terms (Atlanta experienced a tense strike of predominantly black sanitation workers and Detroit suffered a brief summer rampage of young blacks in 1976), such calculations are not perhaps wholly unrealistic. Shortly after Young was elected, for example, a black gang is said to have sent word down to the streets, "Don't embarrass Coleman [Tyson, 1973, p. 681]." Perhaps entirely coincidentally (for this was consistent with nationwide trends), violent crime decreased from high levels in both cities during the mid-1970s. Thus, to white eyes the presence of a black mayor may be seen as a factor in the maintenance of social stability in the marketplace.

All of these various economic factors suggest that the central city economies of Atlanta and Detroit are neither dead nor dying. An economically prosperous downtown plays an especially crucial role in maintaining the economic and psychological health of the city, although a role that must obviously be placed in a qualifying perspective. Business prosperity offers no panacea for the problems with which urban governors must deal. Many central city business enterprises, for example, do not offer jobs for the unskilled or semiskilled urban workforce, and the convention industry of hotels and restaurants often provides little more than underemployment. Nor does the presence of these businesses in their downtown office towers and central city plants magically eliminate the ghetto slums that press against their walls. The substantial taxes they pay are indeed crucial to the maintenance of city services, but nothing guarantees that these revenues will be used for the most needy. There is also no way to ensure that corporate influence and civic energies will be turned to the public uses considered by the community and its government to be the most urgent. In short, any conclusion that major economic forces in Detroit and Atlanta have shown a continued commitment to their respective central cities must be tempered by the knowledge that urban destitution and corporate prosperity have always coexisted in American cities and can continue to do so. In the end, however, what makes the evidence of corporate commitment to Detroit and

Atlanta so important is that the situation could be far worse. No city, governed by blacks or whites, could survive the wholesale flight of its biggest tax producers under the circumstances of the present American political economy.

A STRUCTURAL EXPLANATION FOR THE DIFFUSION OF OPPOSITION

Neither elite acceptance of the legitimacy of black claims to city hall nor the maintenance of commitment to the city necessarily suggests that white political opposition to black mayors will not or cannot form. However, a variety of social and political structural features have apparently made organized and cohesive white opposition an unlikely eventuality in Atlanta and Detroit, at least in the short run. Politics in the two cities manifest a virtually structureless form in which race and personality provide the most meaningful bases of political mobilization. Yet if the racial tie was adequate in the 1970s to hold blacks together as a bloc in political conflict, it was increasingly evident at the same time that mere whiteness offered no such similar basis for effective opposition.

A central problem for potential white opponents of black rule is the nonpartisan structure of local elections in Atlanta and Detroit. In the latter place the strong Wayne County Democratic party organization scrupulously avoids involvement in local politics, particularly since most candidates who compete against one another for elective office in the city are Democrats. Nor does the extensive organizational apparatus of the United Auto Workers union function as an effective surrogate party for local politics in Detroit, for it prefers rather to reserve its energies and resources for national and state contests.

Atlanta, too, lacks a party tradition in local politics. Georgia Persons (1977, p. 5) points out that "neither a Democratic nor Republican party organization plays any formal or significant role in local elections or in the structuring of debates. There are also no strong ward or district political organizations in Atlanta." As a consequence of nonpartisanship, electoral efforts in both cities tend to be structured by candidate organizations, which form around figures with distinctive personalities and racial appeal. White opponents of the black mayors thus have no ongoing institutional structure through which to organize their opposition.

Atlanta and Detroit not only lack local parties but they have no history of machine rule. Machines are particularly energetic competitors when facing displacement or the threat of political defeat.[5] The strength

[5]For an instructive account of the response of a machine to political defeat, see Norwood (1974).

of the resistance of white ethnics to Richard Hatcher in Gary, Indiana, seemed a direct function of the vitality of the machine that long governed the city. White resistance in Chicago to the accession of Wilson Frost, a black city councilman, to the position of acting mayor after Richard Daley's death was swift, cohesive, and effective because it was orchestrated by a machine fighting for survival.

The victories of Maynard Jackson and Coleman Young not only did not displace a party or a machine from power, but they did not displace a particular ethnic group that had exercised a lock on city hall. Black mayors supported by black majorities challenged white rule, but the white communities in both cities had no cohesion based on a common sense of ethnic identity. Displacement brought into play the centrifugal effects of the natural pluralism of the white communities in these cities, a pluralism that reflected generational and ideological differences in the main. Atlanta, notably, has no self-conscious ethnic groups of any consequence at all. With the fall from prominence of politicians like Lester Maddox, its poor white population (the closest analogue to an ethnic group in the city) has no effective spokesmen.

Detroit's ethnic groups, particularly those of eastern European origin, were described repeatedly by white elites as unorganized and lacking in leadership. Other significant ethnic groups there either have little tradition of local political involvement (such as the large "Arab" population) or chose to cooperate through their leadership with the black mayor (such as the Jews). Since the structure of political competition has been defined in those cities in broad racial terms rather than as a function of conflict between blacks and a particular white ethnic group, the sense of displacement within the ethnically diverse (or nonethnic) white communities was not only diffuse but varied. Furthermore, the absence of well-organized ethnic groups meant that no mass level pressures, generated by a sense of ethnic threat, emerged to encourage and support elite resistance.

At least one other aspect of local political structure contributed to the diffusion of opposition and to easing transition. The continued existence of opportunities for white influence through formal roles in city government or on task forces and commissions meant that white elite ambitions and demands in the civic sphere were not entirely blocked. A former Detroit city official (I-15) argued that "the new charter created so many new positions that Young didn't have to get rid of so many of the old guard." But it is important in making this point to understand that there was no evidence to indicate that opportunities for white influence were used to maintain white domination of the cities' politics. No one in or out of government suggested that the agenda of city government had not changed with transition. Nor did anyone hint that either mayor was a tool of white interests or wirepullers. The notion of a covert white power elite had no currency in either city.

In summary, the relatively unstructured character of local politics in Detroit and Atlanta stripped potential opponents of the black mayor of institutional and constituent bases on which to launch opposition. It is, of course, conceivable that even in a situation where such bases did exist, opposition might not form in any event. Overwhelming white support for a black mayor could preclude the use of party, machine, or ethnic group structures to mount resistance and competition. But the point with which we are concerned is whether a structure existed in Detroit and Atlanta in the mid-1970s to facilitate white opposition. To all appearances a facilitating political and social structure did not exist in either place.

CONCLUSION

The culture of accommodation seems to be sustained by displaced white elites in part because it makes good economic sense. To withdraw from the city is attractive only to those with modest, aging capital investments. But for most big firms, existing capital investments coupled with government economic development assistance, including tax abatements, guaranteed loans, public land acquisition for business expansion, job training grants, and public facilities developments, make staying in the city the only sensible economic choice. Thus a good number of firms, even if they wished to escape black city government and a largely black labor force, are virtual prisoners of rational economic imperatives, namely profit motives.

To stay in the city and fight black government is equally irrational, for such an effort risks serious disruption of the marketplace. Contesting black government in a concerted way would be no ordinary political combat, for in some sense it would be challenging the right to govern of a historically deprived and recently victorious majority group. To accept black rule, therefore, is to accept the loss of political control (not a meaningless commodity, as we saw in the previous chapter) to maintain social peace.

The culture of accommodation is not merely and exclusively an economic artifact, however, for it is also sustained by beliefs in the virtues of democratic politics and racial amity. The extent to which these are valued for their own sake and divorced from economic motives is not always clear, to be sure, but it is important to recall that not all of the white elites interviewed for this study were businesspersons with an economic stake in the city. However, even if one wishes to challenge the effectiveness of the moral hold that democratic government and process exercise over those who lose in politics, the fact that certain democratic forms are matters of ingrained habit in America is undeniable. In short, then, environmental factors that emerge from the situational configuration of economic interest, democratic culture, and increasing racial tolerance may explain at one level the maintenance by white elites of the larger culture of accommodation.

9

Political Transitions and the Limits of American Politics

In the previous chapter white acceptance of black political rule was explained as a consequence of environmental factors more or less specific in time and place. The extension of the culture of accommodation to racial politics in Detroit and Atlanta could be understood as a product of increasing racial tolerance in the 1970s, economic interests that white elites both sought to enhance and were in effect bound to maintain, and structural factors peculiar to the two cities. However, it is notable that the outcome of ethnoracial transition in Boston, more than half a century removed from the contemporary events explored in Atlanta and Detroit, was remarkably similar to the modern patterns. Yankee Bostonians responded finally to the establishment of Irish political hegemony by accepting it. Certain of the environmental forces at work in that period were similar to those present in the modern cities and thus might well be supposed to have encouraged a benign transition process. Prejudice toward the Irish was in decline, and Yankee economic investments were firmly in place in the city, at least until the development of the circumferential clean-industry belt on Route 128 after World War II. In addition, the structural features of city government afforded the Yankee minority no real foothold for political opposition.

The similarity of the adjustments to displacement across time, space, and ethnic permutations, however, suggests the distinct possibility that what we witness in all three cases is a product of forces greater than the comparatively particularistic attitudes and economic motives identified as environmental factors. Indeed, the drama of ethnoracial transition, it may be argued, is bound to assume its essentially benign form because of the limits Americans impose on the uses of politics. In some ultimate sense it is these limits that make possible the maintenance of the culture of accommodation.

HISTORICAL CONTINUITIES IN THE TRANSITION PROCESS

To establish the similarities between the process of ethnoracial transition in Boston and in the modern cities, let us reconsider the hypotheses derived from the Boston case study. These were to serve as potential benchmarks as we made our way through the material on the contemporary cities. With a few exceptions it would appear that the patterns observed in Boston have been repeated in Atlanta and Detroit. In some areas, of course, it is simply too early to tell whether Atlanta and Detroit will go the same way as Boston, but there is room for speculation.

TRANSITION AS A GRADUAL PROCESS

In Boston, Yankee and Irish politicians struggled for local power for several decades before the latter were able to establish unbreachable dominance. We cannot say, of course, whether whites as a racial bloc will successfully reassert themselves politically in cities like Atlanta and Detroit, given the short time-line of our observations. Like Boston's Yankee Republicans after the election of O'Brien, whites in black-majority cities have not been sanguine about such a possibility.[1] The Bostonians turned out to be unnecessarily pessimistic, but the white prediction of unbroken long-term black rule appears to be more firmly based on reality. Although Yankee power was undercut to some degree by migration to the suburbs and by higher immigrant fertility rates, these same demographic forces (particularly the former) have been more intense in the modern cities. The possibility still exists, of course, that intraracial political competition in the short-run future will fragment black voting strength so severely that a white candidate could still win

[1]See the pessimistic comments of Atlanta's defeated white mayoral candidate, Milton Farris, in the *Atlanta Constitution* (10 Oct. 1977).

the mayoralty in Atlanta or Detroit, but to all appearances the consolidation of black rule has occurred more swiftly than it did in the case of the Irish.

TRANSITION AS A PEACEFUL PROCESS

The electoral transfer of political power from traditionally dominant groups to formerly subordinate ones was accomplished in all three cities without violent episodes during the campaigns or on election days between members of the groups concerned, without rumors of assassination plots, without incendiary mass meetings, and without the sabotage of governmental processes by members of the displaced group still in a position to carry out such activities. Elites in the new minority embraced a public rhetoric of accommodation rather than resistance, particularly in the modern cases. Despite historic ethnic and racial enmities, displaced elites made no efforts to encourage mass violence or disruption. Such troubles were by no means outlandish possibilities. In Detroit, for example, Mayor Young took pains to vary his route to work through the first year of his initial term to lessen the chances of a preplanned assault on his life.

The peacefulness of the white response in the modern cities, both at the elite and mass levels, stands in obvious contrast to the frequently violent reactions to the civil rights movement in many places in the early 1960s. But it is important to remember that in that earlier period blacks were just beginning to explore their political capabilities and were doing so, of necessity, almost entirely outside the conventional electoral arena. They had not yet made any significant breakthroughs in electoral competition anywhere in the nation.[2] Thus for many in the white community, blacks as political actors in the early 1960s lacked the legitimacy won by successful electoral competition.

CONCESSIONS BY THE NEW MAJORITY

The black mayors were particularly concerned to make clear their intentions to include whites in government and to attend to white interests. If the rhetoric of the displaced elites was accommodating, that of the black victors was generally solicitous and conciliatory. So, too, were the policies of the new mayors, especially with regard to the distribution of appointments and efforts to foster economic growth and develop-

[2]Robin Williams (1977, p. 24) estimates that there were no more than 70 black elected officials in the nation in all elective positions in 1964.

ment. Such actions recall similar concerns on the part of the Irish mayors, who hit upon conservative fiscal policies and selective patronage practices as major strategies by which to reassure Yankee businessmen.

THE PLURALISM OF THE DISPLACED GROUP

In a large and important sense white elites in Atlanta and Detroit developed a high level of consensus on the need to accept and deal openly with black government. But as among Boston's Yankees, there were variations at a more particular level in both strategic and psychological responses to transition. The fragmentation of efforts on behalf of metropolitan reform sharply illustrates the diversity of strategies in one small area. White pluralism was also evident in the level of acceptance of black rule and in the search for ways to establish relations with the black administrations. The relative pluralism of the white communities mainly reflected cleavages based on generational and status differences in Atlanta and ideological ones in Detroit. Many Atlanta elites could not only be distinguished by membership either in the old elite group of the 1950s or in the new young business group but also as heirs of old Atlanta families or newcomers. These cleavages not only affected the possibilities that white elites could work together but also their commitment to adjust to changing times. In Detroit liberal political, labor, and business elites (many of them, in all three sectors, Jews) had staked out strong positions on black rights long before the election of a black mayor was even a remote possibility, thus distinguishing themselves from those political elites and corporate officials who only discovered the urban crisis in the heat of the riot. This is not to say that the modes of adjustment to transition could necessarily be predicted by where an individual was located in regard to these cleavages, for similar responses often crossed cleavage lines. But it suggests that important divisions existed in the white community that militated against the easy development of concerted strategies and monolithic responses.

THE ABSENCE OF RACISM

As in Boston, members of the displaced group were careful not to employ racist language or categories in criticizing the new governors. To some degree strategies of criticism and complaint in Atlanta seemed to depend on the use of surrogates or code categories such as the allegedly poor administrative skills of the mayor, for example (a sign of the elite struggle early in Jackson's first term to come to grips with a political style more or less alien to their experience). In all three cities, however, the achievement of formal political dominance by the former undergroup seemed enough to banish overt racist thinking and criticism from the arsenal of responses. In a sense, power bred respect.

THE DIVERSION OF FEARS

In Boston, Yankee elites focused increasingly on the new immigrants rather than on the Irish as the more profound threat to their culture. As a consequence the Irish came by comparison to look better and better. No such clear diversion occurred in Atlanta and Detroit, however. There was some suggestion in the interview data that many elites saw economic survival as the major crisis facing their cities rather than the rise of black power. Remer Tyson (1975), a Detroit journalist, argued in the *Nation* (p. 237) that the prospects for a national economic depression were so good early in 1975 that Detroiters "had the hate scared out of them." But it is difficult to argue with confidence that such concerns truly had a significant enough diverting effect to enable elites to accept transition more easily, particularly since there were signs of strong economic recovery by 1977.

THE ABSENCE OF WITHDRAWAL

With only a few exceptions Boston Brahmins did not desert their city nor entirely withdraw their civic resources from the public realm (see Whitehill, 1970), although by the 1930s their resources seemed subsidiary to those the Irish had generated through politics. In the modern cities, too, there was no evidence of great withdrawal by elites in the first years of black rule. Indeed, a contrary pattern of commitment and interest seemed to be the rule.

HISTORICAL CONTRASTS IN THE TRANSITION PROCESS

The widespread degree of elite acceptance of black rule and the apparently unorganized, virtually moribund capabilities for white electoral contestation in the 1977 elections seem to indicate that the patterns of accommodation observed in Boston were telescoped in the modern cities. This seemingly swifter process of adjustment may, of course, be an artifact of the short time-line of our observations in Atlanta and Detroit. Yet there are certain historical differences between the Boston situation and that of the modern cities that suggest that the perception of telescoping may not be entirely illusory. Furthermore, even if the transition process has not yet been truly telescoped, it is reasonable on the basis of these differences to predict that benign adjustment will ultimately occur much more swiftly than in the Boston case.

One reason the Yankee struggle to accept the loss of power was so prolonged was the strength and integrity of the New England culture to which the Brahmins were heirs. The rise of the Irish Catholics represented a palpable threat to a well-defined set of values and perspectives. This culture was, moreover, codified in pamphlets, memoirs, and essays. In the more fluid and less intellectually self-conscious society of modern America there is neither an equivalent cultural framework that unites white elites in contradistinction to blacks nor established cultural scribes and arbiters. Although it is true that there may be a distinct black culture, its dimensions are neither unfamiliar to whites nor entirely incompatible with the diversity of white cultural standards. Hence black rule was scarcely seen in the two cities to pose an alien threat to a well-defined white "civilization." The question of cultural survival, so acute for the Bostonians, was never raised in the contemporary cities.

A related difference between the two historical periods is that transition has occurred in the modern cities in a climate of tolerance unknown to Americans at the turn of the century. Anti-Catholic and anti-immigrant doctrines were defining characteristics of the political temper of the decades during which the Irish were consolidating their power in Boston. There existed few larger societal pressures or support for tolerance, as there did during the early years of transition in Atlanta and Detroit.

Another difference between the two periods that led (or may yet lead) to swifter acceptance of rule by the new majority is the present sense that black government may appear to be part of the solution to the "urban crisis." Part of what is wrong in our older cities, many suspect, stems from a failure of confidence in public authority, a concept the Bostonians did not employ. One possible remedy for such a situation is the formal representation in government of those who historically have been left out and are therefore most likely to doubt the legitimacy of public authority. Thus, to the extent that it is believed that black rule can give blacks a greater sense of participation in civic life, help bond them to the polity, help lower crime, provide role models for black youth, and so on, then black government can be seen to be part of the solution rather than part of the problem. That Irish government might possibly have functioned in this same way never occurred to the Brahmins.

TRANSITION IN AMERICA AS A GENERIC PROCESS

The common elements of transition in Boston, Atlanta, and Detroit hint at the presence of underlying regularities in the manner in which Americans are likely to handle certain stages of ethnic political conflict. Once conflicts between ethnic groups work themselves into the electoral arena and competition for the most highly prized offices is pursued on a basis of relative parity, a durable transfer of political control from one ethnic group to another is likely to be peaceful and more or less free of rancor. This is so even when those ethnic groups have a long history of intense animosity and even intergroup violence.

Subsequent efforts to contest transition, once it has occurred, are generally contained within the framework of electoral politics and established governmental institutions. The political displacement of the formerly dominant ethnic group is apparently less likely to galvanize it as a new minority than it is to stimulate its pluralistic tendencies. Indeed, substantial resources controlled by the displaced group eventually become available to the new governors. The black mayors and the Irish mayors were both able to call on elites from the displaced group to commit their resources in a variety of ways conducive to the health of their respective cities.

Responses by both the newly defeated and the newly victorious are governed by a culture of accommodation sustained in some immediate sense by a variety of habitual moral considerations in American politics that we understand as democratic commitments as well as various pragmatic considerations. For example, the need for survival of the new local majority in a larger system in which it is still a minority tempers the behavior of the new victors. The new minority must also survive in its local habitat, particularly if it has economic interests there, and this moderates its responses to the new order.

In the United States such transitions seem to occur primarily in local or urban jurisdictions.[3] It is true that this study has sought neither to discover nor explore instances in which a new ethnic majority emerged at the state level and displaced a traditionally dominant group. But cases

---

[3]The early stages of transition are, perhaps, less likely to be benign in culturally isolated rural or small town settings such as Crystal City, Texas, or Jefferson County, Mississippi. In these less pluralistic, more traditional places, small, cohesive, elite groups, acting out of the larger public eye and in accordance with a set of localized subcultural norms, can more easily respond to displacement in ways that would not fit the patterns observed in the larger cities. Yet even in such places patterns of acceptance and cooperation seem eventually to set in.

of that sort that would have the clarity of the urban transitions do not come readily to mind. This is not to suggest that transition is universally inherently local. In other societies certain kinds of regime changes effect a national transition in ethnic control. This is particularly true where the ruling party is a vehicle for tribal or religious aspirations and is overthrown by a competing party of the same type. One thinks, in addition, of the transition of a different order unfolding in Zimbabwe (Rhodesia). In the United States, however, ethnoracial political transition is a largely local phenomenon. The extraordinary ethnic diversity of the American population militates against the possibility of the emergence of a new ethnic majority in any jurisdiction larger than that of a city. Thus, it has been the cities of America—despite their historic problems of ethnic competition for space, lackluster leadership, corruption, and violence— that have nevertheless been the matrix in which a remarkable degree of ethnic political accommodation has taken place. Cities have not only been the "seat" of ethnic politics, as Robert Lane (1959, p. 239) once observed, but also the vessels in which potentially warring ethnic groups have managed to embrace a common normative framework for peaceful political competition among themselves.

ON THE LIMITS OF AMERICAN POLITICS

In some larger sense it can also be seen that ethnoracial transitions in the United States do not challenge the agreed upon basic purposes and scope of politics. American politics is a process in which certain goods, services, and honors are liable to modest redistribution or extension to the winners but in which certain obligations and established privileges are also regarded as fixed. This seems particularly true in cases concerning the use of political power by groups in office vis-à-vis other groups not in office but that compete in the electoral arena.

Newly victorious ethnic groups, even if they are relatively deprived in American society, do not embark on radical transformations of the social, economic, or institutional structure. They operate within limits, for example, which preserve the broad established patterns and principles of private ownership, taxation, and income distribution. Change is aimed at the extension of goods to the new group (such as the efforts to encourage the growth of black-owned business) or at most to incremental redistribution (such as the opening of government contracts and employment to greater minority participation). The restraints that govern American politics also mean that new victors work more to enlarge the distribution of respect and status than to replace one system of honor with another.

The limits of politics virtually preclude the intrusion of public authority into the sacred or ethnic associational realm. Perhaps most important of all for ethnic groups, political restraint assumes the continued existence of political opposition by any such group that has lost political hegemony. Thus, the fortunes and certain of the most basic interests of ethnic groups that continue to compete in the electoral arena are not threatened in American politics.

These restraints raise some serious, though perhaps not insurmountable, problems (addressed later in this chapter) for a politically victorious group seeking to transform its political power into major social and economic advances. But for the present it is important to understand that all of these limits offer reassurance to displaced groups. If politics were not a limited business, then losers in any political context in the conventional arena could not afford to maintain the norms of accommodation. Too much would be at stake. To put the argument in a realistic context, displaced elites in a system without such limits would have to weigh their accommodation to the new order against the genuine possibilities of wholesale property expropriation (say, through eminent domain condemnations for urban renewal); discriminatory licensing, building code, zoning, and law enforcement; purges of whites from public service; and the withdrawal of all but the most rudimentary police, fire, street repair, and sanitation services from white neighborhoods.

In a different society, the victory of a new ethnic group can mean the incarceration or murder of the formerly dominant group, its dispossession, exile, or the outlawing of its sacred and secular associations. It seems clear, then, that the ability or willingness of displaced groups to adhere to the norms of the culture of accommodation is in the last analysis a function of the limits governing the uses of politics. Environmental forces may act powerfully to encourage acceptance of transition, but without the reassurance of political restraints, economic incentives and cultural pressures for tolerance would have little hold on displaced ethnic groups. Resistance to transition by a united displaced group imbued with a garrison mentality would be the norm in a society without well-established limits. Considered against this not unimaginable context, the capacity of American society to manage ethnic political competition and transition in the electoral arena is a remarkable and important achievement in and of itself.

SOME IMPLICATIONS

The discussion of limits on the use of politics, however, necessarily requires some reconsideration of the significance for blacks or other socioeconomically deprived groups of winning local political power. American electoral politics may permit an extraordinarily high level of ethnic accommodation (seen in world terms), but can politics in such a system be used as a means to social justice? Even assuming the existence of the requisite social technology, it is clear that the limits of politics in America mean that city governments—or indeed any level of government— probably cannot fashion thorough, successful, and speedy policies to deal with the big social justice problems of poverty, discrimination, unemployment, housing, or the quality of urban life.

But this is not to say that government is entirely powerless either, for as the discussion in Chapter 7 made clear, the black mayors have managed to open some significant job opportunities to minorities, change the character of law enforcement, emphasize the importance of neighborhood revival, and so on. What governments in America can but do not always do in other words is to address the big problems incrementally and in piecemeal fashion.

Incrementalism is, of course, a direct product of the limits of politics. Among other things the incremental imperative restricts the resources that government has available to it as well as its scope of authority. Therefore, American governments have traditionally sought to enlist the aid of the private sector in the accomplishment of public ends. Neither the fact of incrementalism nor of government's reliance upon and fostering of private economic activity for public purposes is news, but how these two related factors bear on the achievement of black power in city hall is worth examining further.

INCREMENTALISM

Some might argue that the mere decision to address social justice problems, even incrementally, is no small triumph in itself. Black unemployment, for example, seldom even troubled the city hall of Ivan Allen, much less elicited efforts to deal with it. Nor were such issues as neighborhood preservation, the fostering of black business, or the number of blacks on the police force issues that commanded much attention in official circles up until the mid-1960s and later. But the real question to ask, perhaps, is whether the capture of city hall and its influence in a system of incremental policymaking can lead ultimately to any degree of socioeconomic equity for blacks. Take, for example, the critical problem of jobs. The case of the Irish is instructive on this score, although the comparison to the more favorable situation of modern blacks is, as we shall see, limited to some degree. It is clear that Irish control of city government in Boston and elsewhere led to an enlargement of public sector employment, which the Irish proceeded to fill in disproportionate numbers compared to other ethnic groups (Shannon, 1963, pp. 213–214).[4] Conzen and Lewis (1976) write of Boston that "the Irish used their numerical strength to seize political power and . . . created a buffer to the uncertainties of restricted economic opportunities by greatly expanding the city bureaucracy [p. 25]." Fogelson (1977, p. 124) points out that the Boston police force, nearly two-thirds of which was of Irish descent, tripled in size between 1890 and 1930. Not only did the Irish fill the lower level public jobs in the sanitation and public safety departments, but they also moved heavily into white-collar municipal employment (Erie, 1978).

Although Irish capture of the public job sector may be viewed as an incremental attempt at redistributive politics, it did not immediately improve the group's aggregate economic status (Erie, 1978, p. 289). Nor did Irish political power provide substantial economic mobility through the private job sector, if we interpret Thernstrom's analysis (1973, pp. 131–132) correctly.[5] Finally, Irish political control in Boston did little to promote significant business development in the Irish community, even as late as the 1940s (Shannon, 1963, p. 186). Nevertheless, what the Irish gained by husbanding the public sector was job security for a small but significant portion of their workforce and the creation of a small middle-class core group that had found the avenues of opportunity in the private sector closed.

Although the evidence suggests the absence of a strong direct link between aggregate economic achievement and Irish political power in the early decades of Irish urban ascendancy, it is nevertheless worth noting that the Irish rank today as the most successful white, gentile, ethnic group in America in terms of income and education (Greeley, 1976, pp. 45, 53). The comparison includes both "British" and "American" Protestants. The research exploring the connection between the political power of the Irish over history and their current level of achievement has yet to be done, but it would seem unwarranted to argue that the conjunction is purely coincidental.

Black majority communities that elect their own to city hall can expect at least as much if not more than the Irish won from such incre-

---

[4]See also Clark (1975).

[5]Thernstrom makes no distinction between jobs held in the public and private sectors, but the emphasis appears to be on the latter.

mental efforts to employ and upgrade coethnics in both public and private sector jobs. For one thing the size of the public sector has greatly expanded since the early days of Irish power. Steven Erie (1978, p. 286) calculates that total government employment (federal, state, and local) in big cities in 1900 accounted for slightly less than 4% of the labor force. Local government employment, over which presumably the Irish had greatest control, probably accounted for no more than half the public employees or 2% of the total. In the modern Detroit and Atlanta metropolitan areas (figures for central cities alone are not readily available), total government employment in 1970 was 12% and 15.5%, respectively. Local government employment in the two metropolitan areas amounted to 8% and 7% of the total labor force (U.S. Bureau of the Census, vol. 24, pp. 872–874; vol. 12, pp. 859–861). Percentages for central city municipal employment would probably be higher since suburban local governments tend to be less complex.

Nationally, the white-collar sector of the public labor force is more heavily black today than it was Irish at the turn of the century; also a greater proportion of the total black labor force is engaged in municipal employment than was the case for the Irish (Erie, 1978, pp. 285, 289). Thus, affirmative action in promotion policies is likely to have an effect on aggregate black income in any given city. In addition, locally enforced affirmative action policies now apply to private sector employment in varying degrees of formality, whereas the Irish had no legal basis for bringing leverage to bear on business hiring practices.

In short, black opportunities to secure employment, advancement, and economic mobility through incremental policies aimed at the public and private sectors appear to be considerably more substantial than those the Irish were able to generate. Affirmative action probably cannot do much to combat youth or hard-core unemployment problems among blacks, but it represents nevertheless a major incremental step toward socioeconomic equity in the society.

PUBLIC–PRIVATE PARTNERSHIPS

The limits of politics in America greatly restrict government access to and control of privately owned resources. Therefore, to assemble adequate resources to address certain public problems, government must enter into partnerships with the private sector. Such partnerships are based on the premise that public functions or functions in the public interest may be adequately performed by private sector economic actors at a profit. Venture costs and other subsidies are underwritten by the government. The possibility that black mayors—or indeed any government administration—can effect significant changes in the social order through the partnership principle is based on a number of fragile assumptions. These involve the ability of the city to bear the costs of subsidy, the efficacy of the private sector in fulfilling public needs, and the role of the middle class.

Mayors Jackson and Young had little difficulty forming such partnerships with the business communities of their respective cities. Each was able to call on business elites to commit resources in ways that would presumably eventually create jobs and tax revenues. Eliciting those commitments, however, involves great costs to the city. Public officials, particularly in Detroit, appeared to spend an unusual amount of time, energy, and personal credit in the quest for commitments of business cooperation. Reassuring the business community and generating incentives for its members to stay in the city and invest also involve real money costs. Occasionally these costs are borne by those least able to pay. For example, to demonstrate the tough "business-like" character of their administrations—a way of impressing the business community in order to maintain access to its resources—both mayors drove unusually hard bargains with their public employee unions. In a period in which inflation ran between 7% and 10%, Coleman Young's 4.4% wage increase agreement with the city's civil servants in 1978 was regarded by business as impressive "realism" (I-141). Jackson won business approval for his tough stand against Atlanta's sanitation workers. Toughness and efficiency also reassure the private lending community that purchases city bonds.

Tax rates must be kept low to keep industry, and investment is often gained at the cost of tax abatement or tax increment financing. Gains to the city from such investment are thereby deferred in large part. City tax revenues also subsidize business by helping with industrial land acquisition and supplying the administrative capacity to seek and use federal loan guarantees and job training subsidies. The assumption behind all such activity is that if incentives are attractive enough, private sector actors will act to realize social goals established by the polity.[6] But all this depends on a precise intersection of profit motives and opportunities on the one hand and the public welfare on the other. Unfortunately,

[6] A Committee for Economic Development report issued in early 1978 had the following comments:

> [T]otal private-sector involvement in special training and employment programs for severely disadvantaged groups is much less today than it had been in the late 1960s and early 1970s. . . . Factors contributing to lagging private support for special training and job programs include concern that these activities were imposing an undue burden on the firms' regular profit-making operations [pp. 39–40].

the map to such an intersection is poorly worked out. Can local government in particular provide adequately profitable incentives to ensure private sector cooperation in the achievement of social goals? Or to put the question in another way, can city government afford its dependence on private business for loans, for housing its population, for employing its unemployed, and so on? At what point do the costs of these incentives (which are based on market considerations) outweigh the gains realized through the achievement of social goals (which are determined chiefly in the political arena)? Finally, what guarantees does local government have that the provision of costly incentives to private sector actors will result in the "adequate" realization of social goals? None of these questions is easy to answer, but the record of the private sector as a partner of government in the pursuit of social and economic justice does not provide strong reasons to conclude that the formula has worked well in practice.[7]

Another assumption on which the partnership idea is based is that the role of the middle class in the governing and regeneration of the city is of no more than modest importance. Black-mayor government has been built in places like Atlanta and Detroit on an alliance of the rich and poor, that is, of business and blacks (though educated white voters have tended to support the black mayors). To members of the middle class the incentives to stay in the city or to engage in civic activities may not be readily obvious. If their interests are not directly threatened by black rule, neither are they of central concern. The situation of the white middle class is, ironically, one of benign neglect, a product of the need of the black mayors to maintain access to their prime resource bases— the black community with its votes and the white business sector with its money and jobs. How the middle class will react to its relative exclusion is not yet clear. Certainly its resources—the taxes it pays, its tradition of civic activism—are important to the future of the cities.

In a period in which the notion of partnership is preeminent, however, mayors, black and white, will have little choice but to rely on business–poor coalitions held together by a system of subsidies to attempt to rebuild their cities and employ their jobless. Although the vantage point the mayoralty affords for assembling and even leading such coalitions is important, such partnerships do not hold out the promise of wholesale assaults on poverty, housing, and unemployment. But a radical transformation of politics that resulted in relieving local government

---

[7]"Private sector job creation is not the panacea some proponents claim it to be," is the conclusion of Daniel Hamermesh (1977, p. 34). For an analysis of the pitfalls of the thesis that subsidizing private investment will produce public welfare spillover effects, see Pressman and Wildavsky (1973).

of the need to rely heavily on private sector resources for its economic and, thus, social health, is currently a politically unrealistic and, indeed, uncharted possibility. Furthermore, such a transformation—based as it would have to be on much different patterns of taxation, public ownership and employment, and serious income redistribution in order to address the related problems of cities and urban blacks—would possibly deeply alienate those very whites now committed to the culture of accommodation. It is perhaps ironic, then, that the very achievement that has made possible the peaceful management of mature ethnic competition has also militated against the development of more novel and wide-ranging approaches to the solution of social and economic justice problems in urban America.

List of People Interviewed

DETROIT

Peter Benjamin, journalist
John Boyd, banker
Jerome Cavanagh, former mayor
Bill Ciluffo, city executive office
Avern Cohn, police commissioner
Herman Dooha, New Detroit, Inc.
Larry Doss, New Detroit, Inc.
David Eberhard, city councilman
Bob Fezzey, utilities executive
Sam Fishman, United Auto Workers
Jack Flaherty, corporation executive
John French, banker
Maury Gleisher, public relations
Roman Gribbs, former mayor
Charles Haggler, corporation executive

Note: Many of those interviewed filled other roles besides those listed here.

Dwight Havens, Chamber of Commerce
Marc Ivory, United Auto Workers
Jack Kelley, city councilman
Bernard Klein, former city administrative official
Carl Levin, city councilman
Denise Lewis, city administrative official
Dan Lutzeier, corporation executive
Robert Magill, corporation executive
Maryann Mahaffey, city councilwoman
Fred Matthaei, New Detroit, Inc.
Robert McCabe, Detroit Renaissance
Bruce Miller, Wayne County Democratic Party
William Mitchell, journalist
David Nelson, city executive office
John Nichols, former police chief and former mayoral candidate
Max Pincus, downtown businessman
Jack Pryor, city administrative official
Mel Ravitz, former city councilman and former mayoral candidate
William Ryan, state legislator
Harold Rydholm, corporation executive
Ron Sexton, Detroit Police Officers Association
Michael Smothers, city administrative official
Robert Spencer, Detroit Economic Growth Council
John Steiner, Chamber of Commerce
Bill Stevens, journalist
Robert Surdam, banker
Remer Tyson, journalist
Abe Venable, corporation executive
Leonard Woodcock, United Auto Workers
Ernie Zachary, city administrative official

ATLANTA

Ivan Allen, Jr., former mayor
Ivan Allen III, downtown businessman
Leigh Baier, campaign consultant
Ralph Beck, former downtown businessman
Devin Bent, city administrative official
George Berry, real estate/city administrative official
Don Bradley, neighborhood organizer
Pankey Bradley, city councilwoman

Harold Brockey, downtown businessman
Tarby Bryant, businessman
Helen Bullard, campaign consultant
Bill Calloway, real estate executive
Rodney Cook, former mayoral candidate
Art Cummings, deputy mayor
Brad Currey, banker
Hank Ezell, journalist
Milton Farris, county commissioner
Wyche Fowler, city councilman
Larry Gellerstedt, construction executive
Richard Guthman, city councilman
Tom Hamall, Chamber of Commerce
Allen Hardin, construction executive
Mary Ann Johnson, city administrative official
Willis Johnson, utilities executive
Tom Lowe, county commissioner
Sam Massell, former mayor
Elinor Metzger, League of Women Voters
Jim Minter, journalist
Wynn Montgomery, city administrative official
Reg Murphy, journalist
Jerry Perkins, academic
Hugh Pierce, city councilman
Jocelyn Ross, city administrative official
Bob Royalty, campaign fund-raiser
Virginia Rudder, downtown businesswoman
Carl Sanders, former governor
Bill Shipp, journalist
Clinton Stanford, city administrative official
A. H. Sterne, banker
Dan Sweat, Central Atlanta Progress
Bill Swift, city administrative official
Charles Weltner, former mayoral candidate
Lloyd Whittaker, real estate executive
Sam Williams, real estate executive

Mail Questionnaire Response Data

	Detroit	Atlanta	Totals
Number of firms surveyed (all those employing 1000 or more in SMSA)	67	54	121
Number of firms responding:	37	35	72
Still have facilities in central city	26	26	
Never had facilities in central city	8	9	
Once had facilities in central city but have moved out	3	—	
Number of nonrespondents headquartered in their respective central cities	13	13	
Number of nonrespondents that never had facilities in their respective central cities	17	6	

Total Assessed Property Valuations: Atlanta and Detroit, 1969–1977

	Atlanta	Detroit
	(in billions of dollars)	
1969	1.11	3.77
1970	1.16	3.98
1971	1.18	4.22
1972	1.54	4.35
1973	1.78	4.35
1974	1.81	4.37
1975	1.86	4.27
1976	2.06	4.17
1977	2.08	4.06

Source: Figures supplied by the Atlanta Department of Finance and the Detroit City Assessor's Office.

References

Aberbach, J., and Walker, J. 1973. *Race in the city*. Boston: Little, Brown.

Abrams, R. 1964. *Conservatism in a Progressive era: Massachusetts politics, 1900–1912*. Cambridge: Harvard University Press.

Advisory Commission on Intergovernmental Relations. 1974. *The challenge of local government reorganization*. Washington, D.C., February.

AIA Journal. 1975. Special issue: Report on Atlanta (editor's introduction), 63 (April), 34.

Allen, I., Jr., with Hemphill, P. 1971. *Mayor: Notes on the sixties*. New York: Simon and Schuster.

Allman, T. D. 1978. The urban crisis leaves town. *Harper's*, 257 (December), 41–56.

Andrews, F. M., and Witney, S. B. 1976. *Social indicators of well-being*. New York: Plenum.

Apter, D. 1963. *Ghana in transition*. New York: Atheneum.

Atlanta: The way of all cities? 1975. *The Economist*, March 29, pp. 88–89.

Baltzell, D. 1964. *The Protestant establishment*. New York: Random House.

Banfield, E. 1965. *Big city politics*. New York: Random House.

Banton, M. 1967. *Race relations*. New York: Basic Books.

Barbrook, A. 1973. *God save the commonwealth: An electoral history of Massachusetts*. Amherst: University of Massachusetts Press.

Beale, J. H. 1932. The metropolitan district. In E. Herlihy (Ed.), *Fifty years of Boston*. Boston: Boston Tercentenary Committee.

Blalock, H. 1967. *Toward a theory of minority group relations*. New York: Wiley.

Blodgett, G. 1966. *The gentle reformers: Massachusetts Democrats in the Cleveland era*. Cambridge: Harvard University Press.

209

Brooks, V. W. (Ed.). 1933. *The journal of Gamaliel Bradford, 1883–1932*. Boston: Houghton Mifflin.

Brooks, V. W. (Ed.). 1934. *The letters of Gamaliel Bradford, 1918–1931*. Boston: Houghton Mifflin.

Bryant, T. C. 1975. *The new dynamics of Atlanta*. speech delivered before the Atlanta Hungry Club, April. Mimeographed.

Campbell, A., and Schuman, H. 1969. *Racial attitudes in fifteen American cities*. Ann Arbor, Mich.: Institute for Social Research.

Can Atlanta succeed where America has failed? 1975. *Atlanta Magazine, 15* (June), 40–41, 110–112.

Central Atlanta Progress. 1975. CAP year to date in review. *Re/CAP*, June.

Chamber of Commerce. 1974. *Atlanta*, September.

City government finances. Published annually. Washington, D.C.: U.S. Bureau of the Census.

Clark, T. 1975. The Irish ethic and the spirit of patronage. *Ethnicity, 2*, 305–359.

Cohn, J. 1971. *The conscience of the corporations: Business and urban affairs*. Baltimore: Johns Hopkins Press.

Cole, L. A. 1976. *Blacks in power: A comparative study of black and white elected officials*. Princeton, N.J.: Princeton University Press.

Cole, L. A. 1977. Blacks and ethnic political tolerance. *Polity, 9* (Spring), 302–320.

Committee for Economic Development. 1978. *Jobs for the hard-to-employ: New directions for a public–private partnership*. New York, January.

Connor, W. 1973. The politics of ethnonationalism. *Journal of International Affairs, 27*, 1–21.

Conot, R. 1974. *American odyssey*. New York: William Morrow.

Conzen, M., and Lewis, G. 1976. *A geographical portrait*. Cambridge, Mass.: Ballinger.

Cornwell, E. 1960. Party absorption of ethnic groups: The case of Providence, Rhode Island. *Social Forces, 38* (March), 205–210.

Curley, J. M. 1957. *I'd do it again*. Englewood Cliffs, N.J.: Prentice-Hall.

Dahl, R. 1961. *Who governs?* New Haven, Conn.: Yale University Press.

Dahl, R. 1967. The city in the future of democracy. *American Political Science Review, 61* (December), 953–970.

Dennis, J. 1976. *Public support for the regime in an era of declining confidence in government*. Paper prepared for the Conference on Political Alienation and Public Support, Stanford University.

Detroit Mayor's Office. 1977. *Moving Detroit Forward: A plan for economic revitalization*, June.

Detroit Renaissance. n.d. *A three-year report, 1973–1976*.

Devine, D. 1972. *The political culture of the United States*. Boston: Little, Brown.

Dexter, L. A. 1970. *Elite and specialized interviewing*. Evanston, Ill.: Northwestern University Press.

Dietsch, R. W. 1971. The new, New South: Some progress, more myth. *Nation, 212* (May 17), 615–617.

Edelman, M. 1964. *The symbolic uses of politics*. Urbana: University of Illinois Press.

Edelstein, T. G. 1968. *Strange enthusiasm: A life of Thomas Wentworth Higginson*. New Haven, Conn.: Yale University Press.

Eggler, B. 1970. The old, the new and the never-never South. (Review of *The New South creed* by Paul Gaston). *New Republic, 163* (September 26), 23–25.

Eisinger, P. K. 1976. Ethnic conflict, community-building, and the emergence of ethnic political traditions in the United States. In J. Obler and S. Glass (Eds.), *Urban ethnic conflict: A comparative perspective*. Chapel Hill, N.C.: Institute for Research in Social Science.

Eisinger, P. K. 1978. Ethnic political transition in Boston, 1884–1933: Some lessons for contemporary cities. *Political Science Quarterly, 93*, 217–239.

Enloe, C. H. 1973. *Ethnic conflict and political development.* Boston: Little, Brown.

Equal Employment Opportunity Commission. Published periodically. *Minorities and women in state and local government.* Washington, D.C.

Erie, S. P. 1978. Politics, the public sector and Irish social mobility: San Francisco, 1870–1900. *Western Political Quarterly, 31* (June), 274–289.

Erskine, H. 1973. The polls: Interracial socializing. *Public Opinion Quarterly, 37* (Summer), 283–294.

Fogelson, R. M. 1977. *Big city police.* Cambridge: Harvard University Press.

Friesema, H. P. 1969. Black control of central cities: The hollow prize. *Journal of the American Institute of Planners, 35* (March), 75–83.

Gamson, W. 1968. *Power and discontent.* Homewood, Ill.: Dorsey.

Garraty, J. A. 1953. *Henry Cabot Lodge.* New York: Knopf.

Gaston, P. M. 1970. *The New South creed.* New York: Knopf.

Gelfand, M. I. 1975. *A nation of cities: The federal government and urban America, 1933–1965.* New York: Oxford University Press.

Georgakas, D., and Surkin, M. 1975. *Detroit—I do mind dying: A study in urban revolution.* New York: St. Martin's Press.

Gordon, M. 1961. Assimilation in America: Theory and reality. *Daedalus, 90* (Spring).

Gordon, M. 1964. *Assimilation in American life.* New York: Oxford University Press.

Gordon, M. 1975. Toward a general theory of racial and ethnic group relations. In N. Glazer and D. P. Moynihan (Eds.), *Ethnicity.* Cambridge: Harvard University Press.

Grant, R. 1934. *Fourscore.* Boston: Houghton Mifflin.

Gray, K., and Greenstone, D. 1961. Organized labor in city politics. In E. Banfield (Ed.), *Urban government.* New York: Free Press.

Greeley, A. M. 1972. *That most distressful nation.* Chicago: Quadrangle.

Greeley, A. M. 1976. *Ethnicity, denomination, and inequality.* Beverly Hills, Calif.: Sage.

Greeley, A., and Sheatsley, P. B. 1971. Attitudes toward racial integration. *Scientific American, 225* (December), 13–19.

Green, M. B. 1966. *The problem of Boston.* New York: Norton.

Greenberg, S. B. 1974. *Politics and poverty: Modernization and response in five poor neighborhoods.* New York: Wiley.

Greer, E. 1971. The "liberation" of Gary, Indiana. *Trans-Action, 8* (January), 30–39.

Gribbs, R. 17 March 1976. Interview.

Grodzins, M. 1958. *The metropolitan area as a racial problem.* Pittsburgh: University of Pittsburgh Press.

Hamall, T. n.d. *Crisis/another view.* Mimeographed.

Hamer, A. M. 1973. *Industrial exodus from the central city.* Lexington, Mass.: Lexington.

Hamermesh, D. 1977. Indirect job creation in the private sector: Problems and prospects. Paper prepared for the Brookings Conference on Job Creation, Washington, D.C., April 7–8.

Hammer, P. 1974. *Growth and governance in Atlanta's future.* Extension of remarks before the Atlanta Rotary Club, October 21.

Handlin, O. 1974. *Boston's immigrants.* New York: Atheneum. (Originally published, 1941. Cambridge: Harvard University Press.)

Hanford, C. A. 1932. The government of the city of Boston, 1880–1930. In E. Herlihy (Ed.), *Fifty years of Boston.* Boston: Boston Tercentenary Committee.

Hartshorn, T., Bederman, S., Davis, S., Dever, A., and Pillsbury, R. 1976. *Metropolis in Georgia: Atlanta's rise as a major transaction center.* Cambridge, Mass.: Ballinger.

Harvey, D. 1973. *Social justice and the city*. Baltimore: Johns Hopkins Press.

Hawley, W. 1972. *Blacks and metropolitan governance: The stakes of reform*. Berkeley: Institute of Governmental Studies.

Hein, V. H. 1972. The image of "a city too busy to hate"; Atlanta in the 1960s. *Phylon, 33* (Fall), 205–220.

Hennessy, M. E. 1935. *Four decades of Massachusetts politics, 1890–1935*. Norwood, Mass.: Norwood Press.

Herlihy, E. (Ed.). 1932. Editorial note: The Census of 1930. *Fifty years of Boston*. Boston: Boston Tercentenary Committee.

Hirschman, A. O. 1970. *Exit, voice, and loyalty: Responses to decline in firms, organizations, and states*. Cambridge: Harvard University Press.

Hoar, G. F. 1903. *Autobiography of seventy years*. New York: Scribners.

Hofstadter, R. 1963. *Anti-intellectualism in American life*. New York: Knopf.

Holden, M. 1973. *The white man's burden*. New York: Chandler.

Holland, L. M. 1952. Atlanta pioneers in merger. *National Municipal Review, 41* (April), 182–186.

Huggins, N. I. 1971. *Protestants against poverty: Boston's charities, 1870–1900*. Westport, Conn.: Greenwood.

Hunter, F. 1953. *Community power structure*. Chapel Hill: University of North Carolina Press.

Hutcheson, J. D., Jr. 1973. *Racial attitudes in Atlanta*. Atlanta: Center for Research in Social Change, Emory University.

Hutchinson, E. P. 1956. *Immigrants and their children, 1850–1950*. New York: Wiley.

Huthmacher, J. 1969. *Massachusetts people and politics, 1919–1933*. New York: Atheneum.

International City Management Association. 1974. Police and fire personnel policies in cities over 50,000. *Municipal yearbook*. Washington, D.C.

International City Management Association. Published annually. *Municipal yearbook*. Washington, D.C.

Joint Center for Political Studies. 1976. *Roster of black elected officials*. Washington, D.C.

Jones, M. 1978. Black political empowerment in Atlanta: Myth and reality. *Annals of the American Academy of Political and Social Science, 439* (September), 90–117.

Kantowicz, E. R. 1975. *Polish-American politics in Chicago, 1888–1940*. Chicago: University of Chicago Press.

Kornhauser, A. 1952. *Detroit as the people see it*. Detroit: Wayne University Press.

Kotter, J. P., and Lawrence, P. R. 1974. *Mayors in action*. New York: Wiley.

Kristol, I. 1972. The Negro today is like the immigrant of yesterday. In P. I. Rose (Ed.), *Nation of nations*. New York: Random House.

Kuo, W. H. 1973. Mayoral influence in urban policy making. *American Journal of Sociology, 79* (3), 620–638.

Lane, R. 1959. *Political life*. New York: Free Press.

Lapomarda, V. 1970. Maurice Joseph Tobin: The decline of bossism in Boston. *New England Quarterly, 43* (September), 365–366.

League of Women Voters of Atlanta-Fulton County. 1975. *Facts and figures* (Vol. 48). Atlanta.

Levine, C. H. 1974. *Racial conflict and the American mayor*. Lexington, Mass.: Lexington.

Levine, E. 1966. *The Irish and Irish politicians*. South Bend, Ind.: University of Notre Dame Press.

Lieberson, S. 1963. *Ethnic patterns in American cities*. New York: Free Press.

Liu, B. 1976. *Quality of life indicators in U.S. metropolitan areas: A statistical analysis*. New York: Praeger.

Lockard, Duane. 1959. *New England state politics*. Princeton, N.J.: Princeton University Press.

Lodge, H. C. 1902. *Boston*. New York: Longmans, Green.

Long, M. (Ed.). 1956. *The journal of John D. Long*. Rindge, N.H.: Richard R. Smith.

Long, N. 1958. The local community as an ecology of games. *American Journal of Sociology,* 44 (November), 251–261.

Lowi, T. 1964. *At the pleasure of the mayor*. New York: Free Press.

Mann, A. 1954. *Yankee reformers in an urban age*. Cambridge: Belknap Press of Harvard University Press.

Marquand, J. P. 1937. *The late Georgy Apley*. Boston: Little, Brown.

Mason, P. 1970. *Patterns of dominance*. London: Oxford University Press.

Masotti, L., Hadden, J. K., and Thiessen, V. 1969. The making of the Negro mayors. In L. Ruchelman (Ed.), *Big city mayors*. Bloomington: Indiana University Press.

Massell, S. 18 June 1975. Interview.

McFarland, G. 1975. *Mugwumps, morals and politics, 1884–1920*. Amherst: University of Massachusetts Press.

Mollenkopf, J. M. 1978. The postwar politics of urban development. In W. Tabb and L. Sawers (Eds.), *Marxism and the metropolis*, New York: Oxford Press.

Morison, S. E. 1962. *One boy's Boston*. Boston: Houghton Mifflin.

Moynihan, D. P. and Wilson, J. Q. 1964. Patronage in New York State, 1955–1959, *American Political Science Review, 58* (June), 286–301.

Nation. 1885. *42* (August 13), 124.

Nation. 1886. *43* (December 9), 464.

Neubeck, K. 1974. *Corporate response to the urban crisis*. Lexington, Mass.: Lexington.

New Detroit, Inc. Published annually. *Progress report*.

Nichols, J. 17 March 1976. Interview.

Nordheimer, J. 1974. Toward a new South. *Current, 159* (February), 28–30.

Norton, S., and Howe, M. A. D. (Eds.). 1913. *Letters of Charles Eliot Norton*. Boston: Houghton Mifflin.

Norwood, C. 1974. *About Paterson*. New York: Harper and Row.

Park, R. 1950. *Race and culture*. New York: Free Press.

Parkman, H., Jr. 1932. The city and the state, 1880–1930. In E. Herlihy (Ed.), *Fifty years of Boston*. Boston: Boston Tercentenary Committee.

Perry, B. 1921. *Life and letters of Henry Lee Higginson*. Boston: Atlantic Monthly Press.

Persons, G. 1977. *Black mayoral leadership: Changing issues and shifting coalitions*. Paper delivered at the American Political Science Association meetings, Washington, D.C., September.

Peters, A. J. 1930. Boston—The municipality. In A. B. Hart (Ed.), *Commonwealth history of Massachusetts*, (Vol. 5). New York: States History Company.

Petersen, W. 1975. On the subnations of Western Europe. In N. Glazer and D. P. Moynihan (Eds.), *Ethnicity*. Cambridge: Harvard University Press.

Pettigrew, T. 1972. When a black candidate runs for mayor: Race and voting behavior. In H. Hahn (Ed.), *People and politics in urban society*. Beverly Hills, Calif.: Sage.

Piven, F. F., and Cloward, R. 1967. Black control of cities: Heading it off by metropolitan government. *New Republic, 157* (October 7), 15–19.

Piven, F. F., and Cloward, R. 1968. What chance for black power? *New Republic, 158* (March 30), 19–24.

Poinsett, A. 1970. *Black power, Gary style*. Chicago: Johnson Publishing Co.

Powledge, F. 1975. Atlanta begins to lose its seeming immunity to urban problems. *AIA Journal, 63* (April), 46–51.

Pratt, H. J. 1970. Politics, status, and the organization of Protestant minority group interests. *Polity, 3* (Winter), 222–246.

Pressman, J., and Wildavsky, A. 1973. *Implementation*. Berkeley: University of California Press.

Range, P. R. 1974. Making it in Atlanta: Capital of black-is-beautiful. *New York Times Magazine*, April 7, pp. 28–29; 68–78.

Report of the Mayor's Task Force on City Finances. 1976. Detroit, February 24. Mimeographed.

Report of the National Advisory Commission on Civil Disorders. 1968. New York: Bantam.

Report of the President's Urban and Regional Policy Group. 1978. *A new partnership to conserve America's communities*. Washington, D.C.: Department of Housing and Urban Development, March.

Reuther, V. 1976. *The brothers Reuther*. Boston: Houghton Mifflin.

Roberts, S. V. 1974. He's one of us. *New York Times Magazine*, February 24, pp. 16–35.

Rothchild, D. 1973. *Racial bargaining in independent Kenya*. New York: Oxford University Press.

Ryan, W. 27 April 1976. Interview.

Saveth, E. N. 1948. *American historians and European immigrants, 1875–1925*. New York: Columbia University Press.

Schermerhorn, R. A. 1970. *Comparative ethnic relations: A framework for theory and research*. New York: Random House.

Shannon, W. V. 1963. *The American Irish*. New York: MacMillan.

Shefter, M. 1976. The emergence of the political machine: An alternative view. In M. Lipsky and W. Hawley (Eds.), *Theoretical perspectives on urban politics*. Englewood Cliffs, N.J.: Prentice-Hall.

Shibutani, T., and Kwan, K. M. 1965. *Ethnic stratification*. New York: MacMillan.

Solomon, B. 1956. *Ancestors and immigrants*. Cambridge: Harvard University Press.

Sørensen, A., Taueber, K., and Hollingsworth, L. J., Jr. 1975. Indexes of racial residential segregation for 109 cities in the United States, 1940 to 1970. *Sociological Focus, 8* (April), 125–142.

Stanfield, R. 1978. Federal aid for the cities—Is it a mixed blessing? *National Journal*, June 3, pp. 868–872.

Stark, A. 1973. The challenge and the reality. *Detroit News Magazine* (June 24), 1–7.

Stoffer, D. 1923. Parties in non-partisan Boston. *National Municipal Review, 12* (February), 83–89.

Stone, C. 1976. *Economic growth and neighborhood discontent: System bias in the urban renewal program of Atlanta*. Chapel Hill: University of North Carolina Press.

Storey, M. 1889. *Politics as a duty and as a career*. New York: Putnam's.

Storey, M. 1920. *Problems of today*. Boston: Houghton Mifflin.

Taeuber, K., and Taeuber, A. 1965. *Negroes in cities*. Chicago: Aldine.

Tanner, R. E. S. 1966. European leadership in small communities in Tanganyika prior to independence: A study of conflicting social and political interracial roles. *Race, 7* (January), 289–302.

Thernstrom, S. 1973. *The other Bostonians: Poverty and progress in the American metropolis, 1880–1970*. Cambridge: Harvard University Press.

Tyson, R. 1973. Growing up with Detroit: Coleman Young, street wise and canny. *Nation, 217* (December 24), 681.

Tyson, R. 1975. Mayor Young a year later. *Nation, 220* (March 1), 237–243.

Tyson, R. 1976. Coleman: He's still big daddy. *Detroit Free Press Magazine*, January 4, pp. 7–9.

Underwood, K. 1957. *Protestant and Catholic*. Boston: Beacon Press.

U.S. Bureau of the Census. *Population reports*. Washington, D.C.: U.S. Government Printing Office.

van den Berghe, P. 1967. *Race and racism: A comparative perspective.* New York: Wiley.

Viorst, M. 1975. Black mayor, white power structure. *New Republic, 172* (June 7), 9–11.

Walker, J. 1963. The functions of disunity: Negro leadership in a southern city. *Journal of Negro Education, 32* (Summer), 286–301.

Walton, H. 1972. *Black politics.* Philadelphia: Lippincott.

Ward, D. 1963. *Nineteenth century Boston: A study in the role of antecedent and adjacent conditions in the spatial aspects of urban growth.* Unpublished doctoral dissertation, University of Wisconsin, Madison.

Warner, S. B. 1962. *Streetcar suburbs: The process of growth in Boston, 1870–1900.* Cambridge: Harvard University Press and the MIT Press.

Warner, S. B. 1968. *The private city.* Philadelphia: University of Pennsylvania Press.

Watts, W., and Free, L. A. 1976. *America's hopes and fears–1976.* Washington, D.C.: Potomac Associates.

Weaver, R. C. 1946. *Negro labor.* New York: Harcourt Brace.

Where blacks are moving—and moving up. 1971. *U.S. News and World Report, 70* (March 1), 24–26.

White, M., and White, L. 1962. *The intellectual versus the city.* Cambridge: Harvard University Press and the MIT Press.

Whitehill, W. M. 1970. Who rules here? *New England Quarterly, 43* (September), 434–449.

Widick, B. J. 1972. *Detroit: City of race and class violence.* Chicago: Quadrangle.

Williams, A. W. 1970. *A social history of the greater Boston clubs.* Barre, Mass.: Barre Publishers.

Williams, R., Jr. 1977. *Mutual accommodation: Ethnic conflict and cooperation.* Minneapolis: University of Minnesota Press.

Wilson, J. Q. 1968. *Varieties of police behavior.* Cambridge: Harvard University Press.

World convention dates. 1977. 62 (January).

Young, C. 1965. *Politics in the Congo: Decolonization and independence.* Princeton, N.J.: Princeton University Press.

Zolberg, A. 1964. *One-party government in the Ivory Coast.* Princeton, N.J.: Princeton University Press.

Zolberg, A. 1966. *Creating political order.* Chicago: Rand McNally.

Index

217

Institute for Research on Poverty
Monograph Series

Peter K. Eisinger, *The Politics of Displacement: Racial and Ethnic Transition in Three American Cities.* 1980

Erik Olin Wright, *Class Structure and Income Determination.* 1979

Joel F. Handler, *Social Movements and the Legal System: A Theory of Law Reform and Social Change.* 1979

Duane E. Leigh, *An Analysis of the Determinants of Occupational Upgrading.* 1978

Stanley H. Masters and Irwin Garfinkel, *Estimating the Labor Supply Effects of Income Maintenance Alternatives.* 1978

Irwin Garfinkel and Robert H. Haveman, with the assistance of David Betson, *Earnings Capacity, Poverty, and Inequality.* 1977

Harold W. Watts and Albert Rees, Editors, *The New Jersey Income—Maintenance Experiment, Volume III: Expenditures, Health, and Social Behavior; and the Quality of the Evidence.* 1977

Murray Edelman, *Political Language: Words That Succeed and Policies That Fail.* 1977

Marilyn Moon and Eugene Smolensky, Editors, *Improving Measures of Economic Well-Being.* 1977

Harold W. Watts and Albert Rees, Editors, *The New Jersey Income—Maintenance Experiment, Volume II: Labor-Supply Responses.* 1977

Marilyn Moon, *The Measurement of Economic Welfare: Its Application to the Aged Poor.* 1977

Morgan Reynolds and Eugene Smolensky, *Public Expenditures, Taxes, and the Distribution of Income: The United States, 1950, 1961, 1970.* 1977

Fredrick L. Golladay and Robert H. Haveman, with the assistance of Kevin Hollenbeck, *The Economic Impacts of Tax—Transfer Policy: Regional and Distributional Effects.* 1977

David Kershaw and Jerilyn Fair, *The New Jersey Income-Maintenance Experiment, Volume I: Operations, Surveys, and Administration.* 1976

Peter K. Eisinger, *Patterns of Interracial Politics: Conflict and Cooperation in the City.* 1976

Irene Lurie, Editor, *Integrating Income Maintenance Programs.* 1975

in preparation

Robert H. Haveman and Kevin Hollenbeck, Editors, *Microeconomic Simulation Models for Public Policy Analysis, Volume 1: Distributional Impacts, Volume 2: Sectoral, Regional, and General Equilibrium Models.* 1980